Studies in Systems, Decision and Control

Volume 544

The series "Studies in Systems, Decision and Control" (SSDC) covers both new developments and advances, as well as the state of the art, in the various areas of broadly perceived systems, decision making and control–quickly, up to date and with a high quality. The intent is to cover the theory, applications, and perspectives on the state of the art and future developments relevant to systems, decision making, control, complex processes and related areas, as embedded in the fields of engineering, computer science, physics, economics, social and life sciences, as well as the paradigms and methodologies behind them. The series contains monographs, textbooks, lecture notes and edited volumes in systems, decision making and control spanning the areas of Cyber-Physical Systems, Autonomous Systems, Sensor Networks, Control Systems, Energy Systems, Automotive Systems, Biological Systems, Vehicular Networking and Connected Vehicles, Aerospace Systems, Automation, Manufacturing, Smart Grids, Nonlinear Systems, Power Systems, Robotics, Social Systems, Economic Systems and other. Of particular value to both the contributors and the readership are the short publication timeframe and the world-wide distribution and exposure which enable both a wide and rapid dissemination of research output.

Indexed by SCOPUS, DBLP, WTI Frankfurt eG, zbMATH, SCImago.

All books published in the series are submitted for consideration in Web of Science.

Irina Leonidovna Tarasova ·
Boris Alexandrovich Kulik
Editors

Smart Electromechanical Systems

Mathematical and Software Engineering

Springer

Editors
Irina Leonidovna Tarasova
Saint-Petersburg, Russia

Boris Alexandrovich Kulik
Saint-Petersburg, Russia

ISSN 2198-4182 ISSN 2198-4190 (electronic)
Studies in Systems, Decision and Control
ISBN 978-3-031-64276-0 ISBN 978-3-031-64277-7 (eBook)
https://doi.org/10.1007/978-3-031-64277-7

This Springer imprint is published by the registered company Springer Nature Switzerland AG
The registered company address is: Gewerbestrasse 11, 6330 Cham, Switzerland

Preface

Smart Electromechanical Systems (SEMS) used in Cyber-Physical Systems (CPhS). CPhS have the ability to integrate computing, communication and storage of information, monitoring and controlling of objects in the physical and biological world. The main tasks in the field of theory and practice of CPhS are to ensure the efficiency, reliability and safety of functioning in real time.

SEMS began to be widely used in 2000 in parallel robots or so-called parallel kinematic machines, representing very good capabilities in terms of precision, rigidity and the ability to handle large loads. Currently, SEMS are widely used not only in intelligent robots, but also in other fields: astronomy, machine tool building, medicine, and other fields.

Modern intelligent robots are created based on SEMS modules. A distinctive feature of the new SEMS is the presence of a Central Nervous System (CNS) similar to that of humans. The central nervous system ensures that decisions about appropriate behavior are made in accordance with the goals of SEMS, based on rational knowledge of the environment in which it operates and in accordance with its own technical and mental state.

For the central nervous system, extracting and processing this external information is an integral part of the process of forming their control systems. Moreover, in order for SEMS to operate expediently in a changing and unknown environment, it is necessary to endow them with properties similar to the mental properties of humans. These properties include emotions and temperament, with the help of which the situation of regulating their behavior is assessed. Therefore, it is quite natural to consider the influence of emotions on decision-making in conditions of incomplete information about the environment, especially when SEMS interact with humans to perform complex technological operations.

To date, a variety of mathematical and software tools have accumulated that provide the SEMS central nervous system with the ability to take into account various aspects when interacting with a person.

The purpose of the publication is to introduce the latest advances in the theory and practice of developing mathematical and software SEMS, methods for their design and modeling based on the principles of adaptability, intelligence, parallel kinematics and parallelism in information processing and the formation of control actions.

Topics of primary interest include, but are not limited to the following:

- SEMS information and measurement systems;
- Formation and analysis of databases and knowledge;
- Recognition and classification;
- Decision-making;
- Formation of control actions.

The book is intended for students, scientists and engineers specializing in the field of Smart Electromechanical Systems and robotics and includes many scientific areas, such as the collection, transmission and processing of measurement information, decision theory and optimal control.

We are grateful to many people for their support in writing this book. The list of their names is quite long: Thomas Ditzinger, Janusz Kacprzyk, employees of the Springer publishing house, etc., to whom we are deeply grateful.

On December 24, 2023, the founder and permanent editor of the book series *Smart Electromechanical Systems* Andrey Emel'yanovich Gorodetskiy, Doctor of Technical Sciences, Professor, Honored Worker of Science and Technology of the Russian Federation, Academician of the Russian Academy of Natural Sciences and the Metrological Academy, Honored Inventor of the Russian Federation, head of the laboratory "Smart electromechanical systems" of the Institute for Problems in Mechanical Engineering of the Russian Academy of Sciences, resident of besieged Leningrad passed away.

An outstanding scientist and organizer has died whose scientific works in the field of Smart Electromechanical Systems and control of complex objects and systems, logical-probabilistic and logical-linguistic analysis and synthesis of optimal control systems under conditions of uncertainty are a recognized contribution to world science. His range of interests was very wide and diverse, as evidenced by his recent articles in this book. During his scientific and inventive activity, Andrey Yemel'yanovich has published more than 200 scientific papers, 15 monographs and registered more than 100 patents.

This book is dedicated to the bright memory of Gorodetskiy Andrey Yemel'yanovich.

Saint-Petersburg, Russia
April 2024

Irina Leonidovna Tarasova
Boris Alexandrovich Kulik

Contents

SEMS Information and Measurement Systems

The Accuracy Research of the Exposed Incremental Encoder Based on Image Formation Method for SEMS Applications

Vladimir N. Kuznetsov, Ayur T. Garmaev, Ivan G. Deyneka, and Aleksandr S. Vasilev

Abstract *Problem statement* Linear and angular optoelectronic encoders are among the most precise and high-speed displacement measurement devices for complex control systems. High accuracy and high performance, electromagnetic resistance, and ability to operate in a fast-changing environment determined usage of optical encoders in many SEMS applications. The increasing requirements on SEMS for precise and accurate displacement measurement modules is one of the most demanded topics of modern research. The need to solve problems of accuracy improvement determines the relevance of studying new data processing methods. *Purpose of research* The purpose of the research is theoretical and experimental study of the construction and error estimation of incremental optical encoder based on image formation method. The study examines the influence of design features and data processing algorithms on the measured position error of incremental optical encoder. *Methods* The article discusses the specifics of the incremental optical encoders design based on image formation method. Algorithms for output data processing are considered with respect to real-time applications of adaptive SEMS. Data processing algorithms such as parabola approximation and energy center calculation are presented. Error estimates were obtained for each of the listed algorithms. *Results* Computer modeling and physical evaluation were used for the experimental study. The computer model was implemented and evaluated in Zemax and MATLAB. Physical evaluation was performed on a prototype of linear displacement encoder. Prototype consists of a lens $f' = 3.6$ mm, 5×5 mm beam-splitting cube, sensor OV5640 (pixel size: $1.4\ \mu m \times 1.4\ \mu m$, pixel amount: 2592×1944), FPGA-based

V. N. Kuznetsov (✉) · A. T. Garmaev · I. G. Deyneka · A. S. Vasilev
ITMO University, Saint-Petersburg, Russia
e-mail: vnkuznetcov@itmo.ru

A. T. Garmaev
e-mail: a_garmayev@list.ru

I. G. Deyneka
e-mail: igdeyneka@itmo.ru

A. S. Vasilev
e-mail: a_s_vasilev@itmo.ru

I. L. Tarasova and B. A. Kulik (eds.), *Smart Electromechanical Systems*, Studies in Systems, Decision and Control 544, https://doi.org/10.1007/978-3-031-64277-7_1

computing module with Artix XC7A35T chip. Automated measuring stand based on the Thorlabs LTS300/M linear translator (displacement step from 4 μm, displacement range up to 300 mm) and the XD Laser XD6 LS interferometer (accuracy 0.5 μm/m) was built. For the energy center calculation algorithm 9.85 μm resulting error was obtained at a distance 10 mm. For the parabola approximation algorithm 4.28 μm resulting error was obtained at a distance 10 mm. *Practical significance* The considered algorithms, prototype and developed measuring stand can be used for development of SEMS modules for precise detection of robot joints position. The accuracy research of evaluated displacement measurement methods could be used as a reference for selection of electronic interpolation methods.

Keywords Optical encoder · Optical measurements · High-precision measurements · Grating scale · Linear position sensor · Measurement error · Signal processing

1 Introduction

The development of new production technologies determines the increase in the complexity and speed of positioning operations of high-precision mechanisms [1, 2]. Increasing requirements to the positioning accuracy of the tool relative to workpiece and implementation of accurate feedback is one of the most demanded topics for Smart Electromechanical Systems [3, 4]. Linear displacement encoders are an effective SEMS module for solving such problems [5], providing high speed and measurement accuracy for SEMS control systems.

A linear displacement encoder is a type of measuring instrument designed to determine the spatial position of machine components and mechanisms. Among existing displacement encoder, optical encoders are characterized by the lowest error, highest resolution, and lack of sensitivity to electromagnetic interference [6, 7]. Relative encoders are used in most applications due to the simplicity and relatively low cost of the scale grating [8]. Displacement measurements are performed by detecting the displacement of identical equally spaced strokes. This approach simplifies interpolation and defines the possibility to increase resolution. The quadrature detection method is currently one of the most widespread methods for determining the displacement of a measurement object [9–11]. Thus, the article [11] describes an incremental encoder with a double index grating. The proposed optical scheme of the converter is based on the formation of an image by two diffraction gratings scale with different periods. From the derived intensity distribution equation, the authors conclude that odd harmonics of the third and fifth orders significantly distort the image of the scale. To solve this problem, the authors propose a modified analyzing structure, which is a two-dimensional pattern of strokes. By shifting the upper and lower patterns in phase, it is possible to achieve the suppression of undesirable harmonic components. According to the article, the estimated error of the proposed method is ± 0.18 μm at a distance of 250 mm. The article [12] provides an interpolation method for the quadrature signal of an FPGA-implemented encoder. The method is based on generating phase-shifted sinusoids to calculate an approximated linear signal that is converted into displacement. The theoretical error of the proposed method is 4 nm, the maximum error in the experiment was 0.9 μm per 100 mm of the measuring scale. The study of new methods that improve the metrological characteristics of incremental encoders is a popular topic of modern scientific research [13–15]. These articles propose the design of an incremental encoder based on an imaging principle. The purpose of the article is a theoretical and experimental study of the construction of an exposed incremental optical encoder based on imaging principle and an evaluation of its resulting error. The study examines the influence of design features and data processing algorithms on the position evaluation error of optical encoder.

2 Structure and Principle

The proposed relative optical encoder consists of a coding structure, an image sensor, and a signal processing unit. A coded tape scale was used as a grating scale. In this work, coding tape consisting of a sequence of strokes applied on a polished reflective surface with a period of 40 μm was considered as a grating scale.

The proposed encoder uses an image forming optical system. The optical signal reflected from the grating scale passes through a short-focus lens and is digitized by an image sensor. Based on the received images, the displacement of the grating scale is calculated. The block diagram of the proposed encoder is shown in Fig. 1. Proposed solution reduces the error of position estimation but decrease the maximum movement speed due to the excess number of pixels in the image sensor. Block diagram of the proposed optical encoder is shown in Fig. 1.

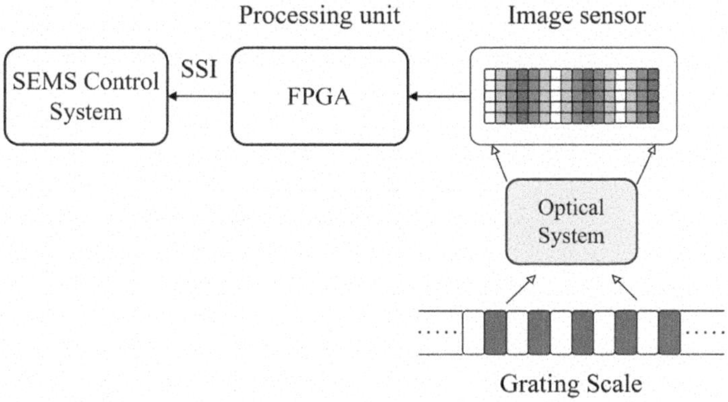

Fig. 1 Block diagram of the optical encoder

To study the resulting error of presented encoder construction, a prototype of an exposed incremental optical encoder was implemented. The light source DFL-5013UWC-2 is directed at a non-polarizing beam splitter cube with geometric size of 5 × 5 × 5 mm, a split ratio of 50/50 and a spectral range of 400–640 nm. Light passing through the cube illuminates the grating scale and reflects from its strokes. The strokes have a size of 20 μm and are located at a distance of 20 μm from each other. Passing the return path through the beam splitting cube, the light is converted by a short-focus lens with focal length f = 3.5 mm and a viewing angle of 90°. The resulting image falls on the OV5640 image sensor as part of the AN5640 evaluation module. The OV5640 is a CMOS matrix with a maximum frame size of 2592 × 1944, a pixel size of 1.4 × 1.4 μm, and the ability to adjust the acquisition area. The image is then transferred to the FPGA, which calculates the displacement of the grating scale. The Alinx AX7035 development board based on the XC7A35T-2FGG484I FPGA was used as an image processing module. This board has a wide range of peripherals for convenient use as a platform for image processing and high-speed data transfer to a SEMS Control Systems or personal computer. The general operating principle of the prototype is shown in Fig. 2.

An example of the resulting image from the described layout is shown in Fig. 3. The one period of detected grating scale fits in approximately 78 pixels of the OV5640.

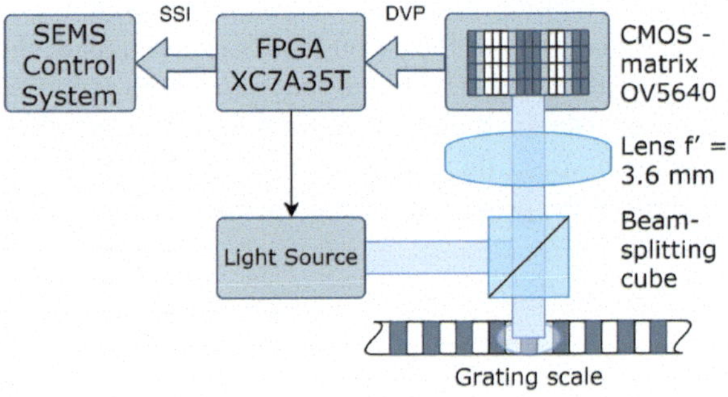

Fig. 2 General operating principle of the incremental optical encoder

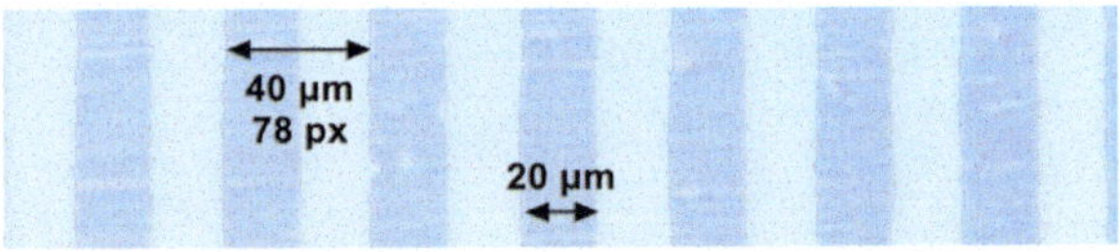

Fig. 3 Image of grating scale

3 Processing Algorithm Development

Two data processing algorithms were developed and studied to determine the best one in terms of implementation complexity and measurement error. An algorithm based on calculating the energy center of each stroke of the coding structure and an algorithm based on approximating the contour of the stroke of the coding structure by a second-degree polynomial (parabola) were considered. A one-dimensional signal obtained by summing all lines of a digital image obtained from a photodetector matrix was used as the initial stroke signal for studying the algorithms.

3.1 Weighted Average Algorithm

For distance computation, an algorithm based on the calculation of the energy center was implemented. From an optical point of view, the energy center of stroke determined the areas with a highest intensity level. They correspond to bright strokes on the grating scale. Stroke detection was implemented by two threshold values to eliminate false triggering. After the stroke is detected, the coordinate of the energy center is calculated. The coordinate of the energy center is the weighted average of the stroke light intensity detected by the pixels of photosensitive element:

$$x' = \frac{\sum_{i=1}^{n} x_i \cdot y_i}{\sum_{i=1}^{n} y_i} \tag{1}$$

where x'—coordinate of the energy center of the stroke, x_i—pixel coordinate, y_i—intensity value of the corresponding pixel, n—the number of pixels in the energy center area.

The evaluation of an energy center begins with determining the maximum signal value in a given area. After the maximum was defined, the algorithm calculates threshold values. Next step is to calculate the weighted average using values between low left and right thresholds by Eq. 1. In the result, the coordinate of the energy center calculated.

Applying this method allows high resolution, which exceeds actual pixel size [16]. The calculation of energy centers could be applied to the entire image obtained from the photodetector. Thus, the output values of the algorithm are a one-dimensional array with the coordinates of the calculated positions of the strokes.

Further calculation of distance comes down to determining the difference of the energy centers coordinates between current and previous frames.

$$X_{diff}^{1}[j] = X_{en.center}[i, j] - X_{en.center}[i-1, j], \tag{2}$$

where X_{diff}—an array consists of difference between energy centers coordinates of current and previous frames, i—frame number, j—energy center number.

When the energy center leaves the field of view of the optical system of the encoder, the movement is calculated using another pair of coordinates. Thus, three differences calculated using Eq. (3):

$$X^2_{diff}[j] = X_{en.center}[i, j-1] - X_{en.center}[i-1, j], \tag{3}$$

$$X^3_{diff}[j] = X_{en.center}[i, j+1] - X_{en.center}[i-1, j], \tag{4}$$

If the difference value between corresponded energy centers coordinates of two adjacent frames exceeds half the period of the structure, it is possible to restore which of these three pairs (Eq. 2–4) corresponds to actual encoder displacement. The resulting displacement value is multiplied by the encoder scale factor to convert the values from pixels to microns.

3.2 Parabola Approximation Algorithm

Another algorithm evaluated as part of the study is based on least squares approximation. The essence of the method lies in calculation of the approximating function coefficients to minimize the root-mean-square error of the deviation of the function from the stroke contour. A polynomial of the second degree is most suitable as a target function for approximation, since it is the simplest even function that closely describes the boundaries of the stroke signal. For a second-degree polynomial, the expression for minimizing the mean square error is as follows:

$$\sigma = \min\left(\sum_{i=1}^{n}\left(y_i - \left(ax_i^2 + bx_i + c\right)\right)^2\right), \tag{5}$$

where n is the number of points in the stroke center area, y_i are the signal values at points x_i.

Then, to find the minimum of the function, the partial derivatives of the expression evaluated:

$$F(a, b, c) = \sum_{i=1}^{n}\left(y_i - \left(ax_i^2 + bx_i + c\right)\right)^2, \tag{6}$$

With respect to a, b and c and equate them to zero. The result will be a system of equations which is used for evaluation of the coefficients of a second-degree polynomial.

$$a\sum_{i=1}^{n} x_i^4 + b\sum_{i=1}^{n} x_i^3 + c\sum_{i=1}^{n} x_i^2 = \sum_{i=1}^{n} x_i^2 y_i$$

$$a\sum_{i=1}^{n}x_i^3 + b\sum_{i=1}^{n}x_i^2 + c\sum_{i=1}^{n}x_i = \sum_{i=1}^{n}x_iy_i$$
$$a\sum_{i=1}^{n}x_i^2 + b\sum_{i=1}^{n}x_i + cn = \sum_{i=1}^{n}y_i \quad (7)$$

By solving the system of equations, the required coefficients of the polynomial a, b, c, are determined, describing the approximating function for a given center of the stroke. Formula 8 determines the coordinates of the stroke center:

$$x' = \frac{-b}{2a}, \quad (8)$$

Solving mentioned system of equations for each stroke in a field of view results in an array of their center coordinates. Further displacement calculation could be performed by using expressions 2–4 in 3.1. The advantage of parabola approximation algorithm lies in an additional noise suppression that occurs as a result of mathematical stroke interpretation.

4 Experiment and Results

4.1 Computer Modeling

To preliminary estimate the error and compare the performance of algorithms in the proposed encoder structure, studies were carried out based on computer modeling in the MATLAB environment. The computer model mathematically described all the main influencing parameters of the encoder nodes, after which an image with the specified parameters was synthesized and the synthesized images were processed using developed algorithms.

The coding scale of an incremental displacement encoder is a ruler with a periodically applied line pattern. To simulate the movement of the encoder, a sequence of images of a coding scale was synthesized. Each frame, the line measure was shifted by the same distance in the selected direction. The synthesis included:

- Harmonic signal generation of a given period, duration, and phase shift;
- Forming a specified image by adding a harmonic signal to a two-dimensional array;
- Noise distortion appliance.

Period of the synthesized structure in length units P = 40 μm or equivalent pixel number $\hat{P} = 77.85$ pixels. The number of pixels n in the image sensor is 1024 pixels. The periodic structure of the coding scale was synthesized based on the following harmonic equation:

Fig. 4 Synthesized image of grating scale with Gaussian noise distribution with $\mu = 0.5, \sigma = 0.06$ parameters

$$I = A\left(\sin(\omega x + \varphi) + \frac{1}{2}\right), \tag{9}$$

where A—amplitude of the signal, ω—angular frequency, φ—phase shift.

The reciprocal of the coding structure period is used as the cyclic frequency of the artificially generated signal. The equation that determines the cyclic frequency of the signal is as follows:

$$\omega = \frac{2\pi}{\hat{P}} = \frac{2\pi}{77.85} = 0.0807\,\frac{\text{rad}}{\text{pixel}}, \tag{10}$$

Simulation of the synthesized data shift was carried out by uniformly incrementing and decrementing the phase of the signal φ. To introduce distorting factors and assess the influence of these factors on the resulting encoder error, an additive noise component was added to the amplitude of synthesized data. The noise component of all main elements of the encoder (image sensor, electronics, light effects, etc.) was approximated by Gaussian noise ($\mu = 0.5, \sigma = 0.06$). The example of a synthesized image shows in Fig. 4.

To model the distorting factors in the image caused by the parameters of the optical system, a model of the optical system of the encoder was synthesized in the Zemax Optic Studio software, with the help of which a synthesized image of the coding scale was formed taking into account optical aberrations. The specified parameters of geometric aberrations for the optical system (a) and resulting image (b) are presented in Fig. 5. The original image was shifted by decentering the optical system and changing the coordinates of the field of view relative to the plane of the object.

The processing errors of data synthesized using the considered computer model are presented in Fig. 6. Designations in the figure: EC—Energy Center, PA—Parabola Approximation.

By adding additive noise with a Gaussian distribution, salt and pepper noise, and exponential noise to the synthesized data, the contamination of the coding structure is simulated. The simulation results show that the considered algorithms are more influenced by noise with a Gaussian distribution, and to a lesser extent by exponential noise. The parabola approximation algorithm generally has a smaller error compared to the energy center algorithm. The largest range of absolute error for the energy center algorithm was 3.86 μm, for the parabola approximation algorithm 2.97 μm.

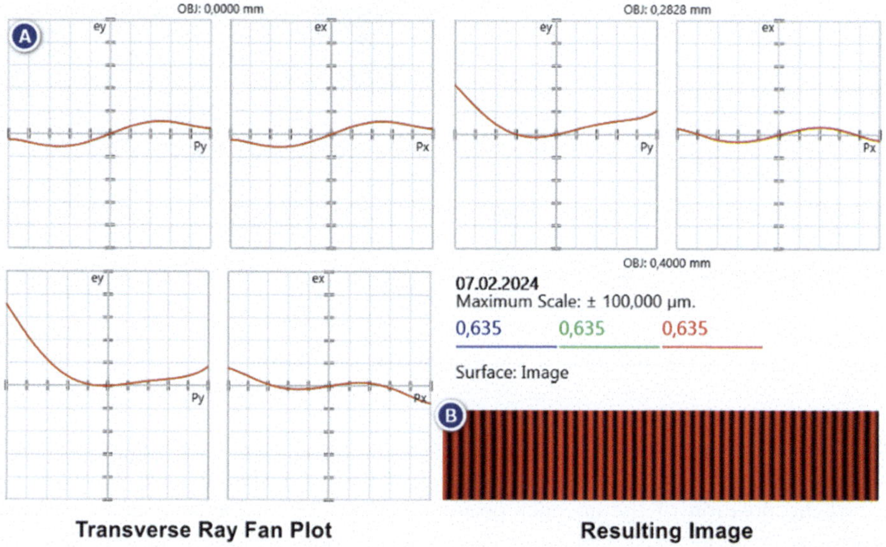

Fig. 5 Geometric aberrations in optical system

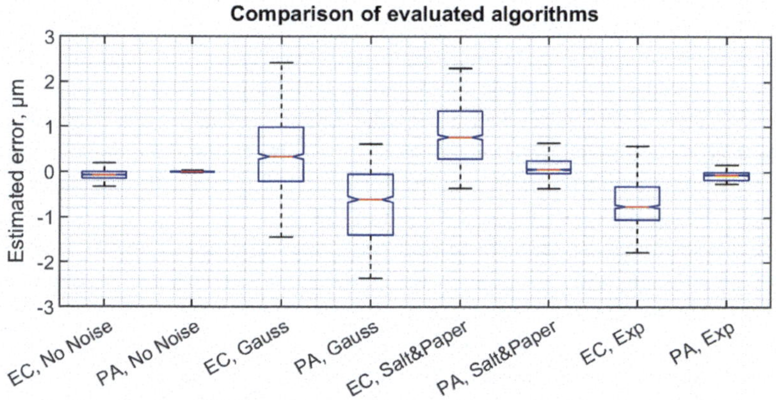

Fig. 6 Algorithm's computational error on synthesized data

4.2 Physical Modeling

To obtain experimental data, the measuring stand was assembled. The diagram of experiment setup (a) and its physical implementation (b) is shown in Fig. 7. The XD Laser XD6 LS (b.1) interferometer was used as a reference measuring instrument. The operating range of the interferometer is 0–45 m with a measurement error of 0.5 µm/m. A Thorlabs LTS300/M (b.2) linear translator is located rigidly on the optical table (b.3). The interferometer reflector (b.4) and the grating scale (b.5) are

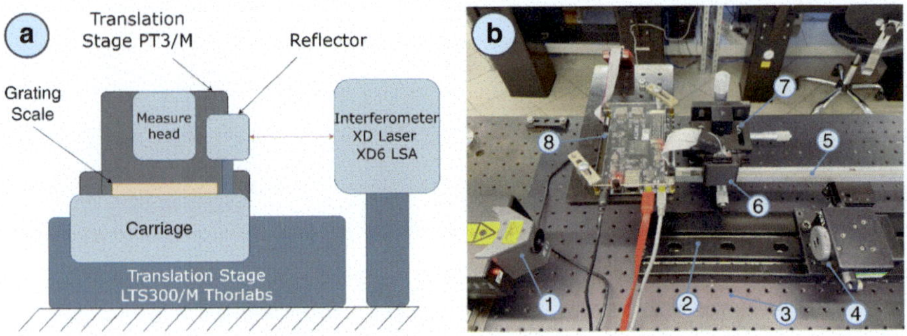

Fig. 7 Experimental setup diagram (**a**), physical implementation (**b**)

fixed on the linear translator carriage. The encoder measuring head (b.6) is mounted on a PT3/M multi-axis micropositioner (b.7). PT3/M translation stage was used to adjust the position of the encoder prototype to achieve a focused image. FPGA processing module (b.8) was connected to the encoder and personal computer. A linear translator, an interferometer and an encoder were connected to a personal computer to implement experiment evaluation and data acquisition.

An experiment was conducted to collect data from the measuring stand. The experiment consisted of moving the grating scale relative to measuring head with defined movement settings. The measurements were carried out with the following parameters: Speed = 0.005 mm/s; acceleration = 0.3 mm/s^2; step length = 0.5 mm, number of steps = 20. After setting up the linear translator, a position reset of XD Laser XD6 LS interferometer was performed. Next, data acquisition from the measuring head and interferometer was performed. A diagram with sequential actions carried out during the experiment is shown in Fig. 8.

The encoder images and interferometer position obtained as a result of the experiment were used as input data for the energy center search and parabola approximation algorithms. The results of data processing are presented in Fig. 9.

Based on the results of the experiment, it is concluded that the noisiness of a real measuring scale, in addition to the studied noise parameters, is characterized by additional sources of error. The error spikes observed in the graph are caused due to an insufficient degree of synchronization between data acquisition from the interferometer and the encoder prototype. The overall difference of derivative of error shows that forward and reverse motion are not equivalent from a data processing point of view. The reason for this may be the presence of backlash in the drive of the linear translator carriage. A graph comparing the statistical characteristics of the error for the considered algorithms is presented in Fig. 10.

Figure 10 shows that the average error value for the energy center algorithm is 1.56 μm, and for the parabola approximation algorithm is 1.48 μm. The magnitude of the error range is from 11.5 μm for the energy center algorithm and 5.4 μm for the parabola approximation algorithm. The conclusion is that the algorithm for

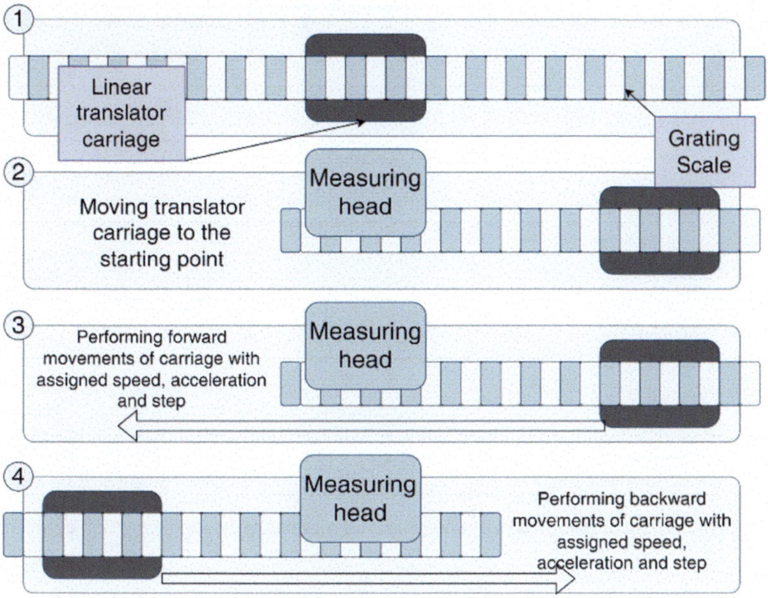

Fig. 8 Sequential stages of the experiment

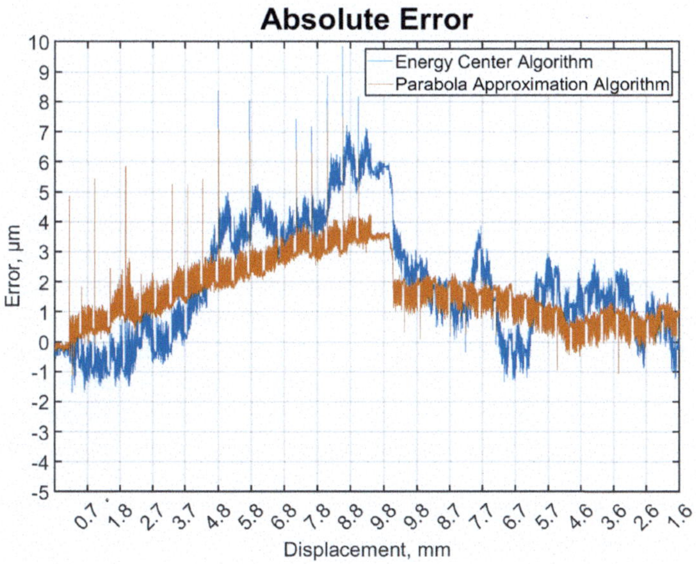

Fig. 9 Algorithm's computational error during the experiment

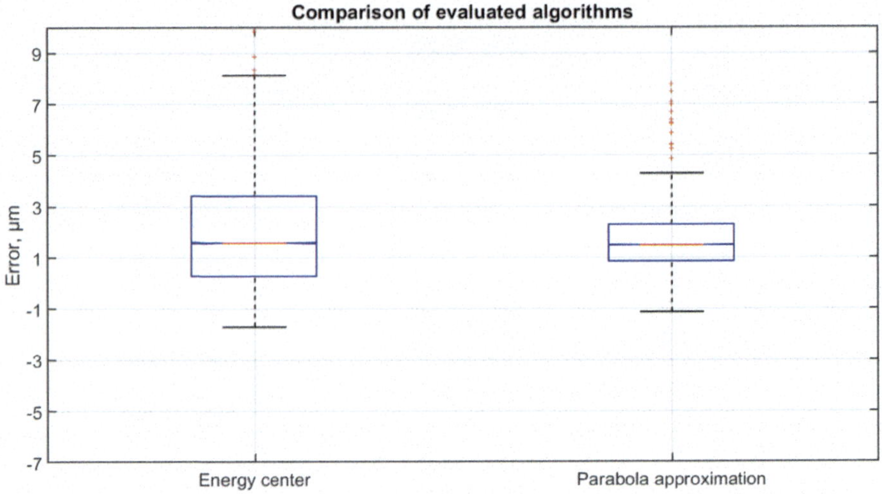

Fig. 10 Error's statistical parameters of energy center and parabola approximation algorithms

approximating the stroke contour with a parabola has a potentially smaller error value.

5 Conclusion

The article presented the process and results of comparing the error of the algorithms based on calculating the energy center and approximating the stroke contour with a parabola for an optical encoder. A computer and physical model of the prototype of the developed optical encoder are presented. The conducted studies showed that the parabola approximation algorithm has a smaller error of stroke center position calculation in comparison with the energy center algorithm. In the case of computer modeling, the magnitude of error for the parabola approximation algorithm was 2.97 µm, which is 23% better than for the energy center algorithm. In the case of physical modeling, the magnitude of error for the parabola approximation algorithm was 5.4 µm, which is 45% better than for the energy center algorithm. Further research will be devoted to determining the influence of the speed of movement of the encoder on a resulting error and the possibility of maintaining the accuracy characteristics of the algorithm at different speeds.

Acknowledgements The research was carried out by the expense of the Russian Science Foundation grant (project No. 23-91-06503).

References

1. Bilancia, P., Schmidt, J., Raffaeli, R., Peruzzini, M., Pellicciari, M.: An overview of industrial robots control and programming approaches. Appl. Sci. **13**, 2582 (2023)
2. Shimizu, Y., Chen, L.C., Kim, D.W., Chen, X., Li, X., Matsukuma, H.: An insight into optical metrology in manufacturing. Meas. Sci. Technol. **32** (2021)
3. Kubela, T., Pochyly, A., Singule, V.: Investigation of position accuracy of industrial robots and online methods for accuracy improvement in machining processes. In: 2015 International Conference on Electrical Drives and Power Electronics (EDPE), pp. 385–388. IEEE (2015)
4. Makulavičius, M., Petkevičius, S., Rožėnė, J., Dzedzickis, A., Bučinskas, V.: Industrial robots in mechanical machining: perspectives and limitations. Robotics **12**, 160 (2023)
5. de Santiago-Pérez, J.J., Osornio-Rios, R.A., de Jesús Romero-Troncoso, R., Herrera-Ruiz, G., Delgado-Rosas, M.: DSP algorithm for the extraction of dynamics parameters in CNC machine tool servomechanisms from an optical incremental encoder. Int. J. Mach. Tools Manuf. **48**, 1318–1334 (2008)
6. Yu, H., et al.: A survey on the grating based optical position encoder. Opt. Laser Technol. **143** (2021)
7. Fleming, A.J.: A review of nanometer resolution position sensors: operation and performance. Sens. Actuatators, A **190**, 106–126 (2013)
8. Bourogaoui, M., Sethom, H.B.A., Belkhodja, I.S.: Speed/position sensor fault tolerant control in adjustable speed drives—a review. ISA Trans. **64**, 269–284 (2016)
9. Tan, K.K., Zhou, H.X., Lee, T.H.: New interpolation method for quadrature encoder signals. IEEE Trans. Instrum. Meas. **51**, 1073–1079 (2002)
10. Cosijns, S.J.A.G., Jansen, M.J., Haitjema, H.: Advanced optical incremental sensors: encoders and interferometers. In: Smart Sensors and MEMs, pp. 245–290. Woodhead Publishing (2018)
11. Ye, G., Liu, H., Ban, Y., Shi, Y., Yin, L., Lu, B.: Development of a reflective optical encoder with submicron accuracy. Opt. Commun. **411**, 126–132 (2018)
12. Ye, G., Wu, Z., Xu, Z., Wang, Y., Shi, Y., Liu, H.: Development of a digital interpolation module for high-resolution sinusoidal encoders. Sens. Actuators, A **285**, 501–510 (2019)
13. Shi, Y., Zhou, Q., Li, X., Ni, K., Wang, X.: Design and testing of a linear encoder capable of measuring absolute distance. Sens. Actuators, A **308**, 111935 (2020)
14. Zhao, G., Ban, Y., Zhang, Z., Shi, Y., Chen, B., Liu, H.: Improving the interpolation accuracy of optical encoders via noise suppression and signal correction. Sens. Actuators, A, 115122 (2024)
15. Ning, Z., Cai, N., Zhao, J., Li, W., Wang, H.: Error compensation for optical encoder based on variational mode decomposition with a coarse-to-fine selection scheme. IEEE Trans. Instrum. Meas. **72**, 1–10 (2023)
16. Lashmanov, O.U., Vasilev, A.S., Vasileva, A.V., Anisimov, A.G., Korotaev, V.V.: High-precision absolute linear encoder based on a standard calibrated scale. Measurement **123**, 226–234 (2018)

Intelligent Support for Selecting Sensors for Information Measurement Systems

Irina A. Koptelova and Lyudmila A. Konovalova

Abstract *Problem statement*: When designing information-measuring and control systems (IMS), the main problem is the informed choice of system elements: sensors, amplifiers, filters, information processing devices and actuators. This problem is due to the fact that there are a very large number of different developments on the modern market. Therefore, the task of selecting elements is to select the most optimal option from a large number of elements. Each element has its own technical characteristics and one of the characteristics may have an advantage over other characteristics. When designing systems, it is necessary to consider a priority list of characteristics for a specific task. The problem of selecting elements for intelligent electromechanical systems SEMS, which can contain a wide variety of elements, devices and subsystems, is especially relevant. When designing a SEMS, the compatibility of elements and devices must be taken into account. Thus, the design of SEMS is quite a complex task and it is obviously possible to solve it using machine methods. The article presents a methodology for the morphological synthesis of IMS, based on the method of hierarchy analysis, which allows selecting the most optimal option based on several criteria. This technique was used in the design of an automated control system for monitoring the geometric parameters of large-sized shells of rotation of the main basic parts of aerospace and power engineering. The IMS contains optical-electronic converters, an electromechanical drive for positioning the converters, which makes it possible to determine their spatial position, at which the control error is minimal. The use of this design technique has shown its effectiveness. However, in this technique, when processing the results of expert assessments, an operation on matrices is used and its influence on the quality of the choice of option is not clear, and the existing technique does not take into account the compatibility of different elements: sensors, information processing devices and electric drives. Therefore, it is necessary to modernize the design methodology taking into account the requirements of SEMS. *Purpose of the study*: to solve the problem of optimal selection of SEMS elements

I. A. Koptelova (✉) · L. A. Konovalova
Volgograd State Technical University, Volgograd, Russia
e-mail: shilina@yandex.ru

L. A. Konovalova
e-mail: eltech@vstu.ru

I. L. Tarasova and B. A. Kulik (eds.), *Smart Electromechanical Systems*, Studies in Systems, Decision and Control 544, https://doi.org/10.1007/978-3-031-64277-7_2

that allow obtaining the maximum possible characteristics of the designed system, it is necessary to conduct research into the influence of all operations of the method on the result of the choice. It is necessary to develop conditions for the compatibility of all elements of the IMS, taking into account the requirements of SEMS. Based on the conducted research, develop a new design methodology. *Results*: A study of all morphological synthesis operations for various tasks was carried out and it was revealed that matrix operations perform the function of averaging the estimates of options. It was proposed to solve the operation of averaging estimates using simpler operations, which makes it possible to reduce the complexity of manual synthesis. Compatibility conditions have been proposed.

Keywords Information-measuring and control systems · SEMS · Heuristic design methods · Hierarchy analysis method · Paired comparisons · Adjacency matrix · Complex priorities

1 Introduction

When developing information-measuring and control systems (IMCS), the main design problem is the informed choice of the main elements of the system: sensors, amplifiers, filters, information processing devices, actuators and positioning devices of measuring transducers. The problem of choice is due to the fact that currently a large number of options for the main elements of systems have been developed. The complexity of the choice problem lies in choosing one option from a set that would satisfy several criteria with the condition of their priority. Each IMS element has its own priority list of technical characteristics, which must be taken into account when choosing an element for a specific task. The problem of selecting elements for intelligent electromechanical systems (SEMS), which may contain mechanical and electromechanical devices and subsystems, is especially relevant. The number of options for electromechanical systems is significantly less than the number of options for sensors and, especially, photodetectors. However, when designing SEMS, it is necessary to take into account the compatibility of ICS elements and electromechanical devices. Thus, designing a SEMS is a rather complex task and it is obviously possible to solve it using heuristic methods. Heuristic methods are widely used in practice to design complex technical systems. Heuristic methods for solving problems do not replace creative search, but organize and direct it, increasing the focus, efficiency and efficiency of design. It should be emphasized that in order to develop heuristic design methods, it is necessary to generalize a large number of solved problems in various fields of technology. Therefore, the greatest effect when using heuristic methods can be obtained by specialists with high creative potential and experience.

From the analysis of existing methods for designing technical systems, the conclusions follow [1] that all heuristic design methods can be divided into two main classes: transformational and morphological methods.

2 Transformation and Morphological Methods

The transformation method of solution is based on a number of principles, namely, at the analysis stage, a plan for solving the synthesis problem is formed, the implementation of which is carried out using various transformations (transformations) of the prototype using heuristic techniques. In this case, the entire cycle of solving a problem (analysis-synthesis) can be repeated many times with various prototypes and heuristic techniques. Transformational methods include the Jones method [2], ARIZ [3] and other methods. In ARIZ, the so-called ideal final result serves as a "guideline model" for the desired solution. It should be noted that transformation methods are not always productive even when the cycle is repeated many times. Purely heuristic operations of obtaining a new technical solution (TS) from its prototype can only be performed by a person, since a computer does not have intuition, cannot predict, build hypotheses, fantasize, etc. Because transformation methods are not formalized and therefore cannot be automated. Attempts to overcome this shortcoming led to the emergence of the morphological method.

The morphological method is based primarily on the combinatorial principle of finding solutions. The procedure of morphological analysis makes it possible to purposefully and systematically include a large number of TPs—analogues of existing TSs, often very distant ones—into morphological sets of solutions.

At the stage of morphological analysis, it is necessary to obtain a morphological set of solutions (MM)—a description of all potentially possible (conceivable) solutions to a given problem, and not a plan for solving the problem. Using the morphological approach, a description of a whole class of systems is constructed on the basis of known TRs. Thus, the morphological approach allows us to obtain many solutions.

A morphological set (MM) can be presented in the form of a morphological table (MT), in which each row represents a classification of the set of systems under study according to some essential characteristic. Therefore, the morphological table is the result of multidimensional classification. Each line of the MT contains not only known, but also all acceptable values of the characteristics, and therefore such a MT contains a large number of fundamentally new variants of TR. Moreover, some of them may not be of practical interest, i.e. morphological tables are redundant. At the stage of morphological synthesis, the most optimal ones are selected from a large number of new TR contained in MT. Morphological synthesis methods are heuristic search methods that use a limited search of options. The heuristic nature of morphological synthesis methods lies in the use of techniques that make it possible to find the required option according to the conditions of the problem in a number of steps that are one to two dozen orders of magnitude less than with an exhaustive search. For example, with the help of MT, variants of TR of optical-electronic measuring instruments with a cardinal number of 21,027 were obtained [1]. The number of features in this MT was more than 50, and the number of values ranged from 2 to 10.

Morphological method of system design. Morphological synthesis of systems with a constant structure must be carried out in the following sequence: precise formulation of the problem posed, selection of functionally complete blocks of the system, comparison of each selected block with the values of its implementation and bringing them into a table and selection from the MM of the most optimal set of blocks that would provide the best values of operational system characteristics.

To solve this problem, it is advisable to use the priority setting method, which is based on the task of choosing a leader [4].

Let us consider the method of morphological synthesis using the example of designing an optical-electronic system for controlling the technological process of producing large-sized parts in a heated state 6 [5, 6]. Let's take a closer look at the example of choosing photodetectors, which have a large number of options and with a large number of parameters and characteristics [7].

For the morphological synthesis of the optical-electronic system, the main parameters of photodetectors were selected: k_1—receiver temperature, k_2—time constant, k_3—supply voltage, k_4—long-wave boundary, k_5—dark resistance, k_6—dark current, k_7—specific detection ability.

The square adjacency matrix is constructed as follows: the number of the option is written in the columns and rows, and at the intersection—the coefficients (1.5; 1.0; 0.5), indicating which option is preferable. To calculate β_j (Table 1) each row (k_1–k_7) in the matrix is multiplied by a column vector Σ. Normalized values, i.e. relative values $\beta_j{}'$ are obtained by β_j dividing by $\sum_{i=1}^{m} \beta_i$. Relative values $\beta_j{}'$ represent a list of priority parameters of the selected elements or criteria significance coefficients.

The above method allows for sequential selection of elements for each specific parameter. At the same time, a certain type of photodetector can provide the required system parameter with varying degrees of ease, while some parameters cannot be provided by this photodetector at all. For example, operation in the long-wavelength range is relatively easily provided by bolometers and photoresistors, but photomultipliers are practically inoperable in the long-wavelength range. The conditions for high-speed operation of an optical-electronic device are also provided by various photodetectors with varying degrees of ease. Therefore, a method is needed that would take into account not only the priorities of each parameter, but also the degree

Table 1 Adjacency matrix of photodetector criteria

Criterion index	k_1	k_2	k_3	k_4	k_5	k_6	k_7	$\sum_{i=1}^{n} k_i$	β_j	$\beta_j{}'$
k_1	1.0	1.5	1.5	0.5	1.5	1.5	1.5	9.0	59.0	0.187
k_2	0.5	1.0	1.5	0.5	1.5	1.5	1.5	8.0	50.5	0.160
k_3	0.5	0.5	1.0	0.5	1.5	1.5	1.5	7.0	43.0	0.136
k_4	1.5	1.5	1.5	1.0	1.5	1.5	1.5	10.0	68.5	0.217
k_5	0.5	0.5	0.5	0.5	1.0	0.5	0.5	4.0	26.5	0.084
k_6	0.5	0.5	0.5	0.5	1.5	1.0	0.5	5.0	31.0	0.098
k_7	0.5	0.5	0.5	0.5	1.5	1.5	1.0	6.0	36.5	0.115
Total	–	–	–	–	–	–	–	–	315.0	1

of ease of providing this parameter with the type of photodetector, i.e. it is necessary to introduce weighting coefficients for each type of photodetector.

3 Photodetector Selection Technique

Let's consider the method of choosing a photodetector taking into account the particular criteria k_1, k_2,... k_7 from five options X_1, X_2, X_3, X_4, X_5. Each option corresponds to a type of photodetector, namely X_1—photoresistor, X_2—photodiode, X_3—bolometer, X_4—photomultiplier, X_5—photomagnetic receiver. Assessment in accordance with the above method is carried out using complex priorities B_{iKOM} according to the formula [4].

$$B_{iKOM} = \sum_{i=1}^{n} \beta_j' B_{ij}' \tag{1}$$

where $\beta_j{}'$—is the relative priority (significance) of the *j*th criterion; $B_{ij}{}'$—relative priority of the *i*th option according to the *j*th criterion; *n*—number of criteria ($n = 7$).

At the first stage, square matrices are constructed for each of the criteria (Table 1). The number of the option is written in the columns and rows, and at the intersection—the coefficients (1.5; 1.0; 0.5), indicating which option of the criterion is preferable.

Priorities of criteria.

The next stage involves the sequential determination of the absolute priorities Bij of options, and then the relative ones B_{ij}, which are calculated in fractions of one (Table 2). To calculate $B_{ij}{}'$, each row (X_1–X_5) in the matrix is multiplied by a column vector $\sum_{i=1}^{n} k_i$. Normalized values, i.e. relative $B_{ij}{}'$ are obtained by dividing B_{ij} by $\sum_{i=1}^{m} B_{ij}$. According to this rule, priorities are calculated for Tables 3, 4, 5 and 6 of the criterion $B_{ij}{}'$.

Element priorities according to a given criterion.

The priorities of photodetectors for each criterion are determined in the same way.

Table 2 Adjacency matrix of photodetectors according to the first criterion

Variant index	Adjacency matrix by 1st criterion					$\sum_{j=1}^{m} X_j$	B_{i1}	$B_{i1}{}'$
	X_1	X_2	X_3	X_4	X_5			
X_1	1.0	0.5	1.0	0.5	0.5	3.5	16.00	0.136
X_2	1.5	1.0	1.5	1.0	1.0	6.0	28.50	0.242
X_3	1.0	0.5	1.0	0.5	0.5	3.5	16.00	0.136
X_4	1.5	1.0	1.5	1.0	1.0	6.0	28.50	0.242
X_5	1.5	1.0	1.5	1.0	1.0	6.0	28.50	0.242
Total	–	–	–	–	–	–	117.50	1

Table 3 Relative assessments of criteria, alternatives and complex indicators of the importance of photodetectors

Significance criteria $\beta_j{'}$	Relative priority for each option				
	X_1	X_2	X_3	X_4	X_5
0.187	0.136	0.242	0.136	0.242	0.242
0.160	0.182	0.241	0.137	0.197	0.241
0.136	0.172	0.267	0.172	0.120	0.267
0.217	0.239	0.191	0.295	0.121	0.152
0.084	0.196	0.196	0.196	0.119	0.290
0.098	0.121	0.173	0.238	0.294	0.173
0.115	0.152	0.239	0.191	0.295	0.121
Comp. prior	0.176	0.223	0.197	0.193	0.209

Table 4 Adjacency matrix and relative priorities of photodetector criteria

Index criterion	k_1	k_2	k_3	k_4	k_5	k_6	k_7	$\sum_{i=1}^{n} k_i$	*kcp*	β_j	$\beta_j{'}$
k_1	1.0	1.5	1.5	0.5	1.5	1.5	1.5	9.0	1.286	11.571	0.218
k_2	0.5	1.0	1.5	0.5	1.5	1.5	1.5	8.0	1.143	9.143	0.173
k_3	0.5	0.5	1.0	0.5	1.5	1.5	1.5	7.0	1.000	7.000	0.132
k_4	1.5	1.5	1.5	1.0	1.5	1.5	1.5	10.0	1.429	14.286	0.270
k_5	0.5	0.5	0.5	0.5	1.0	0.5	0.5	4.0	0.571	2.286	0.043
k_6	0.5	0.5	0.5	0.5	1.5	1.0	0.5	5.0	0.714	3.571	0.067
k_7	0.5	0.5	0.5	0.5	1.5	1.5	1.0	6.0	0.857	5.143	0.097
Total	–	–	–	–	–	–	–	–	53	1	

Table 5 Matrix of paired comparisons (on a ratio scale) and relative priorities of photodetector criteria

Index criterion	k_1	k_2	k_3	k_4	k_5	k_6	k_7	$\sum_{i=1}^{n} k_i$	*kcp*	β_j	$\beta_j{'}$
k_1	1	2	2	0.5	2	2	2	11.5	1.643	18.893	0.233
k_2	0.5	1	2	0.5	2	2	2	10	1.429	14.286	0.176
k_3	0.5	0.5	1	0.5	2	2	2	8.5	1.214	10.321	0.127
k_4	2	2	2	1	2	2	2	13	1.857	24.143	0.297
k_5	0.5	0.5	0.5	0.5	1	0.5	0.5	4	0.571	2.286	0.028
k_6	0.5	0.5	0.5	0.5	2	1	0.5	5.5	0.786	4.321	0.053
k_7	0.5	0.5	0.5	0.5	2	2	1	7	1.000	7.000	0.086
Total	–	–	–	–	–	–	–			81.25	

Table 6 Matrix of paired comparisons of criteria (on a 9-point scale of relations) and relative priorities of criteria

Index criterion	k_1	k_2	k_3	k_4	k_5	k_6	k_7	$\sum_{i=1}^{n} k_i$	kcp	β_j	$\beta_j{}'$
k_1	1.000	7.000	7.000	0.111	5.0	3.000	2.0	25.11	3.587	90.08	0.333
k_2	0.143	1.000	7.000	0.143	4.0	2.000	2.0	16.28	2.327	37.89	0.140
k_3	0.143	0.143	1.000	0.200	3.0	3.000	2.0	9.486	1.355	12.85	0.047
k_4	9.000	7.000	5.000	1.000	2.0	2.000	2.0	28.00	4.000	112.00	0.414
k_5	0.200	0.250	0.333	0.500	1.0	0.333	0.5	3.11	0.445	1.39	0.005
k_6	0.333	0.500	0.333	0.500	3.0	1.000	0.25	5.91	0.845	5.00	0.018
k_7	0.500	0.500	0.500	0.500	4.0	2.000	1.00	9.00	1.286	11.57	0.043
Total	–	–	–	–	–	–	–			270.78	

Then a complex indicator (priority) is calculated for each of the $B_{i\text{ком}}$ options (Table 3), which is defined as the sum of the products of the relative priorities of the object by the relative priorities of the criteria. The last line indicates complex priorities, which were obtained by summing the products of the elements of the first and subsequent columns of the table (formula (1)) (Fig. 1).

The option that received the highest $B_{i\text{ком}}$ value can be considered the best of all the others. In our case it is $B_{2\text{ком}} = 0.223$. Therefore, device number 2 (option X_2-photodiode) is considered the best.

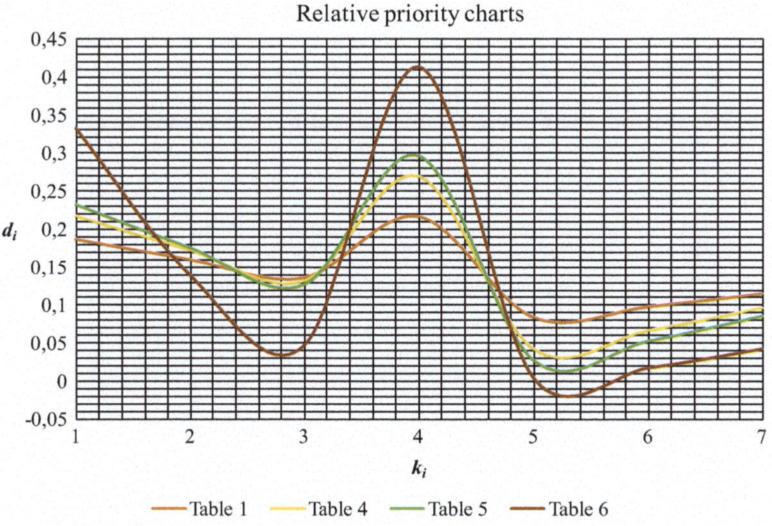

Fig. 1 Relative priorities of different options

The above methodology has shown high efficiency in the design of systems in various fields of technology. One of the problems with this relatively simple technique is the operation of multiplying a row and a column. From the analysis of this operation, its main function was not revealed. Therefore, it was proposed to replace this operation by multiplying the sum of the row by its average value.

The fourth method used a nine-point relationship scale with multiplicative comparison (Table 6).

4 Conclusion

From the conducted studies of different ranking methods, it follows that the ranking order of the modernized methods in comparison with the main method has not changed. However, the second technique using the average value and additive comparison requires fewer operations and therefore can be used without the use of special programs. The morphological synthesis method allows you to create the most optimal UES design from a set of known components and elements than using traditional design methods. It should be noted that formalized design methods are the basis for the creation of computer expert systems and computer-aided design systems.

References

1. Odrin, V.M.: Method of Morphological Analysis of Technical Systems, p. 312. VNIIPI, Moscow (1989)
2. Jones, J.K.: Design Methods: Translated from English, p. 326. Mir, Moscow (1986)
3. Altshuller, G.S.: Invention Algorithm, p. 296. Moscow Worker, Moscow (1973)
4. Moiseeva, N.K., Karpunin, M.G.: Fundamentals of the Theory and Practice of Functional Cost Analysis, p. 192. Higher School, Moscow (1988)
5. Shilin, A.N.: Design of adaptive optical-electronic devices for monitoring the shaping processes of large-sized heated parts. Control Diagn. **7**, 14–20 (2001)
6. Miroshnikov, M.M.: Theoretical Foundations of Optical-Electronic Devices, p. 696. Mechanical Engineering (1983)
7. Ishanin, G.G.: Receivers of Radiation from Optical and Optoelectronic Devices, p. 175. Mechanical Engineering (1986)

Intelligent Information-Measuring System for Relay Protection of Overhead Power Lines with Adaptation to External Factors

Aleksey A. Shilin, Pavel V. Dikarev, and Danila N. Avdeyuk

Abstract *Problem statement*: In the electric power industry, relay protection is designed to prevent accidents in the power supply system. In the event of a short circuit in the power supply systems, relay protection is activated, which, using electromechanical switches, cuts the power off the consumed electrical network. The relay is activated when the emergency current exceeds a previously set threshold value—the setpoint. The main problem in relay protection is the justified choice of the setpoint value, which depends on a large number of different factors: climatic, geographical and the type of vegetation along the power line route. Currently, the value of the setpoint is mainly determined approximately on the basis of operating experience, which, due to the mismatch of the setpoint with the emergency-dangerous current value, leads to the relay protection being triggered in the absence of an accident or the protection not being triggered in the event of an accident. Some problems of increasing the reliability of protection operation were solved, automatic correction of electrical circuit parameters was introduced into the relay protection system based on measured values of climatic factors: temperature, humidity, precipitation level, wind. This made it possible to increase the reliability of relay triggering, however, during the tests, unknown reasons for the unreliable triggering of the relay has been revealed. Based on the analysis of environmental influences, the following factors has been identified: terrain, vegetation and swampiness. It should be noted that all these factors are unformalized and therefore these problems can be solved with the help of an intelligent information-measuring relay protection system controlled by electromechanical switches of overhead power lines, which together with electromechanical switches is SEMS. *Purpose of the study*: to solve the problem of increasing the reliability of the operation of electromechanical relay protection, which reduces the risk and consequences of accidents in power supply systems, it is proposed to use the Elcut 6.6 software package, which allows simulating the electromagnetic field of an electric network and an artificial neural network (ANN), which allows to increase the level of adaptation of the SEMS setpoint value to changes in external factors. *Results*: A study has been carried out using the Elcut 6.6 software package, based

A. A. Shilin (✉) · P. V. Dikarev · D. N. Avdeyuk
Volgograd State Technical University, Volgograd, Russia
e-mail: eltech@vstu.ru

I. L. Tarasova and B. A. Kulik (eds.), *Smart Electromechanical Systems*, Studies in Systems, Decision and Control 544, https://doi.org/10.1007/978-3-031-64277-7_3

on which a method was developed for determining the capacitance of line wires relative to the ground surface based on the given parameters of grass vegetation and trees, as well as the geometric dimensions of ravines and squares of swampiness. Based on the obtained wire capacitance values, the total resistance of the short circuit along sections of the route was correcting. To implement the carried out research, the intelligent information-measuring system has been developed using SEMS relay protection of overhead power lines. An algorithm for controlling an electromechanical switch has been developed based on a neural network. *Practical significance*: The developed intelligent electromechanical system SEMS during implementation will increase the reliability of relay protection triggering and reduce the risk and consequences of accidents in power supply systems. In addition, the results of the study can be used when designing to determine the topology of networks with the highest possible reliability of electricity transmission.

Keywords Relay protection · Setpoint of triggering · Artificial neural networks · SEMS · Factors influencing line resistance · Capacitance of line wires relative to the ground surface

1 Introduction

One of the main problems in the electric power industry is the reliability of power supply to consumers. The most unreliable elements of the electrical power system are overhead power lines. This is due to the fact that the lines are located in different geographical zones and are exposed to difficult climatic factors—rain and snow, wind and ice loads. To prevent emergency conditions, relay protection is used, which, in the event of a short circuit in the power supply systems, is activated and, using electromechanical switches, de-energizes the consumed electrical network. The relay is triggering when the emergency current exceeds a previously established threshold value—the setpoint. The main problem in relay protection is the justified choice of the setpoint value, which depends on a large number of different factors. In many countries around the world, the problem of increasing reliability is being solved by introducing Smart Grids. Smart Grids are a complex of technical means that automatically identify the weakest and most emergency-dangerous sections of the network, and then change the characteristics and layout of the network in order to prevent accidents and reduce losses [1]. In addition, the smart grid must be self-diagnostic and self-healing and incorporate advanced sensor, communication and control technologies to improve the efficiency of energy transmission and distribution. Thus, a smart grid is an automatically balancing and self-monitoring energy system, capable of receiving energy from any source (coal, sun, wind) and converting it into a final product for consumers (heat, light, warm water) with minimal human participation. From the above definitions and tasks to be solved, it follows that the main technical means (Smart Grid) should be an intelligent information-measuring relay protection system controlled by electromechanical switches of overhead power lines, which together with electromechanical switches is SEMS [2].

The main problem in relay protection is the justified choice of the setting current value, which depends on a large number of different factors: climatic, geographical and the type of vegetation along the power line route. Currently, the setpoint value is determined approximately, which leads to the relay protection being triggered in the absence of an accident or the protection not being triggered in the event of an accident. Some problems of increasing the reliability of protection operation were solved earlier by introducing into the relay protection system automatic correction of electrical circuit parameters based on measured values of climatic factors.

2 Analysis of External Factors of Overhead Power Lines

Based on the carried out analysis of external factors influencing the reliability of relay protection operation, factors were identified that were not previously taken into account when choosing the triggering setpoint, namely the terrain and the nature of plantings, water spills, the level of snow precipitation, as well as the electrification of the air around the line due to the triboelectric effect. To automatically correct the setpoint, a mathematical model was built to determine the total resistance of the wire, taking into account the influence of external factors. For the technical implementation of automatic correction of the setpoint, the information-measuring system must contain a set of sensors: current, air temperature, wind speed and direction, soil conductivity, as well as a reflectometer. Using the Elcut software package, a simulation of the process of the influence of unformalized values of external factors on the triggering setpoint of the relay protection was carried out, from which it follows that in the event of a single-phase ground fault (SGF), in order to take into account the line capacity, the information-measuring system must also contain the following sensors: wire sag, size and weight of ice deposits, temperature of the wire, level of precipitation (snow, water and ice) and deep depressions in the terrain, volumetric air charge. In addition, a two-coordinate description of the terrain along which the line route passes is necessary.

The listed factors are difficult to formalize, and fuzzy logic and an artificial neural network were used to solve this problem. The work carried out a study of the functions of the influence of physical quantities of external factors on the total capacitance of overhead power line wires when determining the single-phase ground fault current in networks with an isolated neutral. Based on the study of the functions of the influence of external factors on the total capacity of the overhead power line wire quantitative estimates of the influence of each factor were obtained, which allows one to reasonably, depending on the requirements, choose the value of the factor for correcting the relay protection setpoint. For the technical implementation of the research results, an information-measuring system was developed to control the relay protection setpoint, which implements processing the values of external factors and determines the adjusted value of the relay protection setpoint based on an artificial

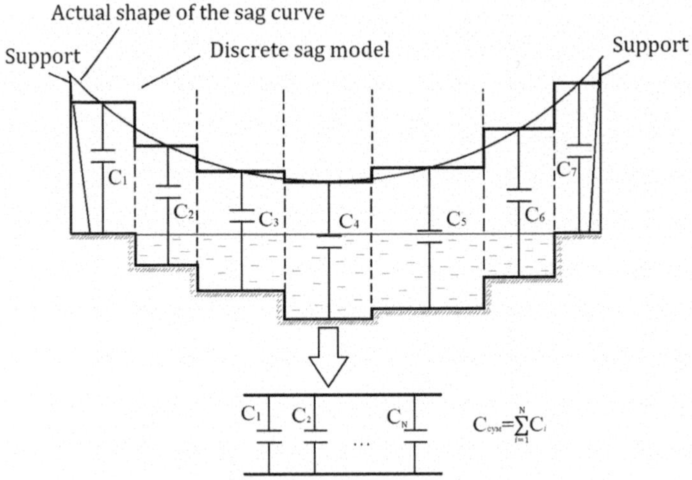

Fig. 1 Scheme for calculating the capacity of a wire passing through a ravine filled to the brim with snow/water/ice/other mixtures with water

neural network and fuzzy logic. For the information-measuring system for controlling the relay protection setpoint, a training technique for an artificial neural network has been developed, which makes it possible to more accurately determine the relay protection setpoint for correct triggering due to the influence of external factors on the parameters of the overhead power line [3].

To calculate the parameters and characteristics of SEMS, studies were carried out on the influence of newly discovered influencing factors on the reliability of relay protection triggering: wire sag, terrain, shapes of hills and depressions, the degree of their filling with snow, water, ice [4], forest cover of the route, icy deposits, volumetric charge of the air gap surrounding the wire. Figure 1 shows a diagram for calculating the total wire capacitance on the analyzed span with the influence of factors.

Using the Elcut software package, the tasks of the functions of the influence of some factors were simulated, not taking into account which causes an error in calculating the wire capacitance relative to the surface of the earth, namely, a ravine and sagging of the wire can lead to an error in determining the wire capacitance of 9.21%; not taking into account the layer of dense vegetation can leads to an error of 1.95%; not taking into account several factors (Fig. 2a, b) can lead to an error in capacity calculation of 17.56%.

Figure 2b shows the geometric model of the problem in Elcut, simulating the conditions for passing the line route shown in Fig. 1a.

The appearance of an ice coupling [5] 5 mm thick already leads to an error exceeding 20%.

During snowstorms and sandstorms, there is an accumulation of volumetric charge in the air space surrounding the overhead line wires. This phenomenon is explained by the triboelectric effect, which in the environment can manifest itself in the process of intense friction of snow particles and ice crystals during strong downstream wind

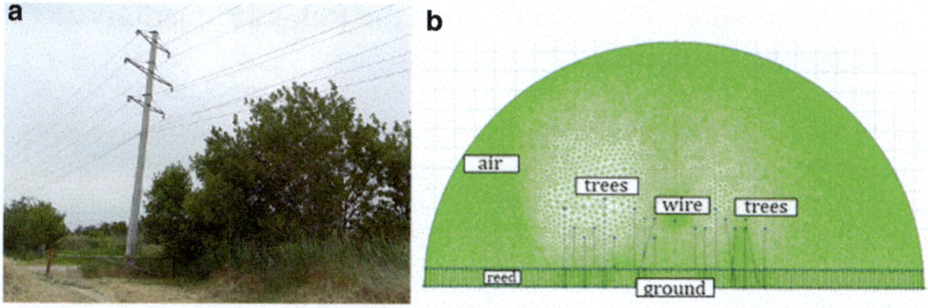

Fig. 2 **a** passage of the overhead line route through a forested area (reeds, forest thickets), **b** geometric model of this problem in Elcut

flows. Moreover, the volumetric charge accumulated in the air, especially in the presence of freshly fallen snow, can reach 10–8 C/m^3.

Non-accounting such electrification of the environment around the wire can lead to an error exceeding 79%. As follows from this, the triboelectric effect is a significant factor of error when determining the capacitance of a wire relative to ground.

Based on the conducted research, an algorithm has been developed for application the digital shadow (digital twin) of the electrical network in the Matlab Simulink environment to emulate the transient processes of real overhead power lines, which allows filling the database for training the neural network to adjust the relay protection setpoint taking into account changes in line parameters due to the influence of external factors.

To make a reasonable choice of SEMS element base and an algorithm for their compatibility, a methodology was developed based on the hierarchy analysis method, which, based on the technique of paired comparisons of various options, allows you to select elements that satisfy several criteria, taking into account predetermined weighting coefficients and compatibility conditions of the designed system. This method [6] solves the pressing problem of selecting the most optimal element from a large number of existing options. This is currently a problem because there are a large number of different options available in the sensor market and the main challenge is choosing one that satisfies the requirements. If there is no sensor that meets the requirements of the system being designed, then such a sensor is developed according to technical specifications. Using this method, sensors has been selected to monitor physical quantities—temperature of the wire and air, speed and direction of air masses, soil conductivity, electrification of the air around the line wire during snowstorms due to the triboelectric effect, line current load, water level and terrain near power lines, as well as the type and amount of precipitation.

3 Development of SEMS to Control the Relay Protection Triggering Setpoint

Figure 3 shows a functional block diagram of the developed relay protection setpoint control SEMS. The information-measuring system consists of functional blocks, the interaction between which is carried out by a microprocessor unit. All modules are powered from the built-in power supply unit (PSU).

The microprocessor unit (MU) receives analog current and voltage signals from the unit of transformers (CVTU), as well as the value of the distance to the accident site from the reflectometer (R) and signals from external sensors, namely the earth conductivity sensor (ECS), wind speed and direction sensor (WSDS), air temperature sensor (ATS), volumetric air charge sensor (VACS), type and amount of precipitation sensor (TAPS) and water level sensor (WAS). Via a common bus in digital form, the MU receives the state of the control buttons from the indication and control unit (ICU) and requests from the upper-level system from the communication unit (CU); control and signaling signals are issued to the relay control unit (RCU), LED signaling signals to the indication and control block and responses to the request of the upper-level system to the communication block. Reception and output of external signals is carried out in accordance with the algorithms of the application and system software of the microprocessor unit.

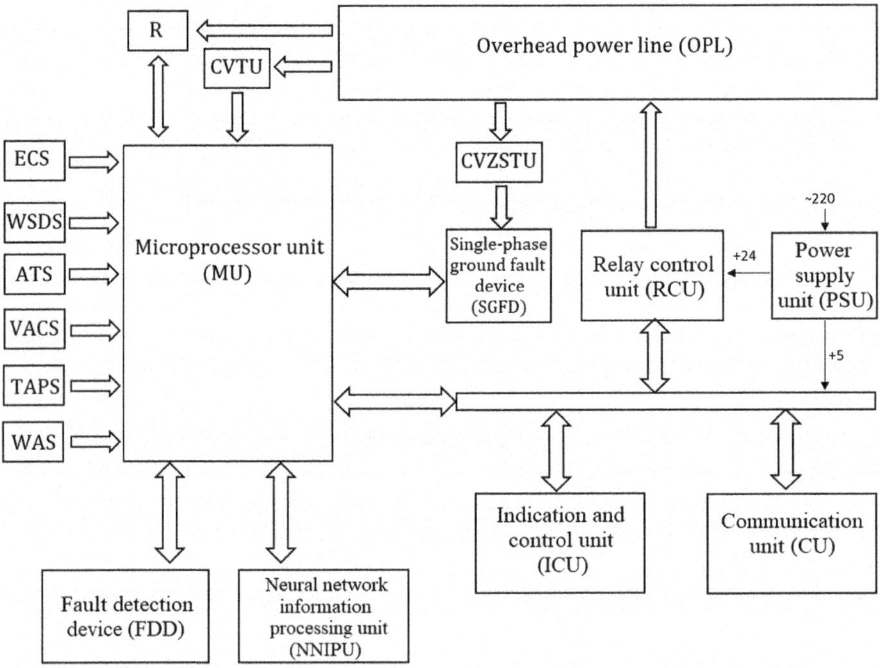

Fig. 3 Functional block diagram of the relay protection setpoint control SEMS

The microprocessor unit (MU) receives analog current and voltage signals from the unit of transformers (CVTU), as well as the value of the distance to the accident site from the reflectometer (R) and signals from external sensors, namely the earth conductivity sensor (ECS), wind speed and direction sensor (WSDS), air temperature sensor (ATS), volumetric air charge sensor (VACS), type and amount of precipitation sensor (TAPS) and water level sensor (WAS). Via a common bus in digital form, the MU receives the state of the control buttons from the indication and control unit (ICU) and requests from the upper-level system from the communication unit (CU); control and signaling signals are issued to the relay control unit (RCU), LED signaling signals to the indication and control block and responses to the request of the upper-level system to the communication block. Reception and output of external signals is carried out in accordance with the algorithms of the application and system software of the microprocessor unit.

The MU transmits the current and voltage values in the line to a fault detection device (FDD), which determines the presence of a fault in the line based on the received values using fuzzy logic and transmits the result (presence of damage and its type) back to the MU. In the presence of multiphase damage, the MU, using functional dependencies, calculates the primary parameters of the line, the phase velocity V_Ph. Using the obtained value of the phase velocity V_Ph, the reflectometer calculates the distance to the accident site and transmits it to the MU, which calculates the deviations of the primary parameters of the line and transmits these values to the neural network information processing unit (NNIPU). The NNIPU determines the actual setpoint value of triggering using a neural network based on the deviation values of the primary parameters and transmits the setpoint value to the MU, which compares it with the actual value of the current in the line, makes a decision to disconnect the line and, if necessary, transmits a control signal to the relay control unit (RCU) via a common bus. In the event of a single-phase ground fault (SGF), the MU transmits values from sensors, transformers and reflectometer to the single-phase ground fault devise (SGFD). The SGFD also receives values from the current and voltage zero-sequence transformers unit (CVZSTU). Based on the data obtained, using a neural network, the SGF setpoint is determined and compared with the current total capacitive current of the line. The result of the operation of the SGFD is transmitted to the MU, from where a control signal is transmitted to the RCU for triggering the relay protection "on a signal", as well as the value of the distance to the accident site at the CU.

The power supply unit has several types of built-in protections (cut-off, overload, etc.). The transformers units provides galvanic isolation and scaling of input signals from the measuring transformers of current, voltage, zero-sequence current and zero-sequence voltage to the level required by the analog–digital part of the MU. The relay control unit carries out control the output relays, which are equipped with one switching contact. The microprocessor unit processes incoming signals and generates control action via the bus. The ICU is used to control the device and display operational information and settings.

4 Conclusion

Protection systems against SGF in networks with an isolated or compensated neutral are specific and the setting current in such protections is determined differently than as is customary for protection against multiphase short circuits. Moreover, such systems should not disconnect the damaged line, but respond “on a signal” [7].

The proposed neuro-computer system for protection against SGF can be characterized as an intelligent agent due to its compliance with such parameters as adaptability, learning ability, and the ability for self-development. By means of a digital shadow following the real electrical network, updating of training examples for the neural processor as a processing subagent is achieved. The neural network is systematically retrained at a given frequency, which allows us to talk about the adaptability of the protection system in the process of interaction with the environment.

Acknowledgements The research was funded by the Russian Science Foundation (project No. 24-29-00430, https://rscf.ru/en/project/24-29-00430/).

References

1. Evdokunin, G.A.: On the principles of constructing relay protection against single-phase ground faults in networks 6–35 kV. In: Tikhodeeva, N.N. (ed.) Overvoltages and Reliability of Electrical Equipment Operation. Materials of the International Scientific and Technical Conference. Budapest, November 14–18, 2005, Issue 4. PEIPK, St. Petersburg (2006)
2. Kamel, T.S.: Application of artificial intelligence systems in distance protection of power lines. In: Kamel, T.S., Moustafa Hassan, M.A., El-Morshedy, A. (eds.) Modern Directions of Development of Relay Protection Systems and Automation of Power Systems: Proceedings of the 2nd International Scientific and Technical Conference, Moscow, September 07–10, 2009, pp. 30–38. NP Scientific and Engineering Information Agency, Moscow (2009)
3. Zilberman, V.A.: Relejnaya zashchita seti sobstvennyh nuzhd elektrostancij (Relay Protection of Power Plant Auxiliary Networks). M.: Energoatomizdat (1992)
4. Dielektricheskie svojstva vody i l’da. Sposoby ochistki i svojstva vody. (Dielektricheskie svojstva vody i l’da. Sposoby ochistki i svojstva vody). Available at: https://www.o8ode.ru/article/krie/Dielectric_properties_of_water_and_ice. Accessed 28 Feb, 2024
5. Khakimzyanov, E.F.: Matematicheskaya model’ zaderzhki i zatuhaniya vysokochastotnyh signalov v liniyah elektroperedachi s gololyodnymi obrazovaniyami (Mathematical Model of Delay and Attenuation of High-frequency Signals in Power Lines with Ice Formations). Energy Tatarstan. (2), 24–28 (2011)
6. Shilin, A.N.: Teoriya prinyatiya reshenij v proektirovanii informacionno-izmeritel’noj tekhniki (Theory of Decision Making in the Design of Information and Measuring Equipment). In: Shilin, A.N., Koptelova, I.A. (eds) VSTU. Volgograd, p. 128 (2012)
7. Shabad, M.A.: Zashchita ot odnofaznyh zamykanij na zemlyu v setyah 6–35 kV (Protection Against Single-phase Earth Faults in 6–35 kV networks). In: Shabad, M.A. (ed). M.: NTF Energoprogress, p. 64 (2007)

Intelligent Information-Measuring System for Controlling the Operating Mode of Hydroelectric Power Plants

Aleksandr N. Shilin, Lyudmila A. Konovalova, and Muluken A. Bogale

Abstract *Problem statement* One of the main problems in the operation of hydraulic structures of hydroelectric power stations is maintaining the water level in the reservoir. The inflow and consumption of water resources depends on many factors: meteorological and hydrological conditions, volumes of water resource consumption. The balance of water resources in the reservoir is monitored using sensors that measure water inflow and flow. The main problem of maintaining the water level in a reservoir over a long period is the optimal distribution of water resources. Exceeding a certain retaining value of the water level can lead to its overflowing over the dam, and a drop in the water level below this disrupts the normal operation of the hydroelectric power station. The flow of water resources is controlled by gates with electromechanical drives. Typically, the management strategy of a hydroelectric power plant is determined based on previous meteorological and hydrological conditions and information about the delay and current water level in the reservoir. It should be noted that accurately predicting the water level in a reservoir is a difficult task, since a hydraulic structure is an extended object that is affected by a large number of different random and unpredictable factors. Therefore, to solve this problem, it is proposed to use an intelligent information and measurement system (IMS) for controlling electromechanical drives of reservoir gates. Thus, the IMS for controlling electric gate drives for the tasks being solved is SEMS. *Purpose of the study* to solve the problem of optimal distribution of water resources in a reservoir, allowing to ensure normal operation of the hydroelectric power station and meet the needs of other consumers, taking into account the flow of water into the reservoir, it is proposed to use an artificial neural network (ANN). ANN allows you to solve the problem of predicting the flow of water into a reservoir from rivers, often flowing over a large area with different weather conditions. *Results* to implement the proposed solution, an intelligent information and measurement system for controlling electromechanical drives of reservoir gates (SEMS) has been developed. The structure of the neural network

A. N. Shilin (✉) · L. A. Konovalova · M. A. Bogale
Volgograd State Technical University, Volgograd, Russia
e-mail: eltech@vstu.ru

L. A. Konovalova
e-mail: eltech@vstu.ru

I. L. Tarasova and B. A. Kulik (eds.), *Smart Electromechanical Systems*, Studies in Systems, Decision and Control 544, https://doi.org/10.1007/978-3-031-64277-7_4

and the methodology for its training have been developed, taking into account the statistical database, geographical and weather factors. *Practical value* The developed intelligent information and measurement system for controlling electromechanical drives of reservoir gates (SEMS), which, when implemented, will reduce fluctuations in the water level in the reservoir and, therefore, ensure the normal operation of the hydroelectric power station. In addition, this system can be used to control water balance in other hydrological structures, such as irrigation systems.

Keywords Information and measuring system · Artificial intelligence system · SEMS · Hydroelectric power station · Reservoir · Control of water level in the reservoir · Artificial neural network

1 Introduction

One of the global problems of life support for the population on the planet is the provision of water resources. Reservoirs of hydroelectric power plants (HPPs) play a major role in the storage and distribution of water resources, and therefore, for the normal functioning of HPPs and water supply to all consumers, it is necessary to maintain the water level in the reservoir. To maintain the water level in the reservoir, it is necessary to control the balance of water resources, namely: inflow and flow, which depends on many factors: meteorological and hydrological conditions and the volume of consumption of water resources and electricity by the population, industry, agriculture and environmental requirements. The main problem of maintaining the water level in a reservoir over a long period is the optimal distribution of water resources. The management strategy for a hydroelectric power plant is usually developed based on previous meteorological and hydrological conditions and information about the retention and current water level in the reservoir. The balance of water resources in the reservoir is monitored using sensors that measure water inflow and flow. Depending on the water level and forecasts of its arrival, decisions are made on early release of water, which avoids the risk of flooding in downstream areas. The water level in the reservoir determines the power output of the hydroelectric power plant, and fluctuations in the water level lead to changes in the power output of the generators. Therefore, accurately predicting the water level in a hydroelectric power station's reservoir is very important when planning power generation. Correct and effective planning of energy consumption allows us to ensure a stable supply of electricity to consumers. It should be noted that accurately predicting the water level in a reservoir is a difficult task, since a hydraulic structure is an extended object that is affected by a large number of different random and unpredictable factors. Therefore, to solve this problem, it is necessary to use an appropriate forecasting method. As such a method, it is advisable to use neural networks, which are successfully used in solving similar problems. An artificial neural network (ANN) [1] reproduces the functioning of a simplified mathematical model of a biological neural network in software or hardware. The use of neural networks has wide applications in various fields, since their main properties are nonlinearity and generality. This means that a neural network can produce good results even for input samples that are different from the training samples. Thus, an artificial neural network (ANN) will solve the problem of predicting the flow of water into a reservoir from rivers that often flow over a large area with different weather conditions. It is obvious that the solution to such a complex problem is possible with the help of an information-measuring system containing sensors of physical quantities, a neural network and a control computer. The flow of water resources should be controlled by gates with electromechanical drives. Thus, in terms of the tasks to be solved, the control information system in conjunction with electric gate drives is SEMS.

2 Information-Measuring System for Controlling Electromechanical Drives of Hydroelectric Reservoir Gates

To solve this important problem, an information-measuring system for controlling electromechanical drives of hydroelectric reservoir gates was developed, which contains two groups of sensors for the main factors influencing the water level: climatic and water. Climatic sensors: air temperature 1, precipitation level 2, snow cover level 3, soil moisture content 4, freezing depth 5, which are installed on the banks of the river. Water sensors: water evaporation 6, located on the surface of the water, water level 7, inflow flow 8 and water consumption 9, located under water. All sensors, climate and water, are placed on both sides of the dam. Sensors 1–9 are connected to a control computer 13, containing a neural processor 14, which generates a forecast of the inflow and water level in the reservoir and, based on this information, plans water consumption for consumers, taking into account their priorities and flood risk, and an interface 15. The computer 13 carries out control by the spillway gate 10 through electric drive 12. The IIS for controlling the water level in the hydroelectric power station reservoir is implemented as follows.

Sensors IIS 1–9 form a data array for transmission to a computer 13 with a neural processor 14, which simulates an artificial neural network trained to forecast water level and reservoir flow with a given horizon of N days. The neural processor 14 performs a forecast, actually approximating the functions of changes in two quantities—the water level in the reservoir and the flow rate of the reservoir—based on information received through data transmission channels from sensors 1–9, as well as information accumulated by the computer 13 about the dynamics of changes in electrical loads in the power system (Fig. 1).

The neural processor 14 operates in the modes of training and interrogating the trained ANN to predict the values of the mentioned two quantities and retrain the ANN after comparing its predicted value with the actually recorded value. In the training and retraining (re-training) modes, the computer 13 processes arrays of information using a program, comparing input data (from sensors 1–9 and electrical

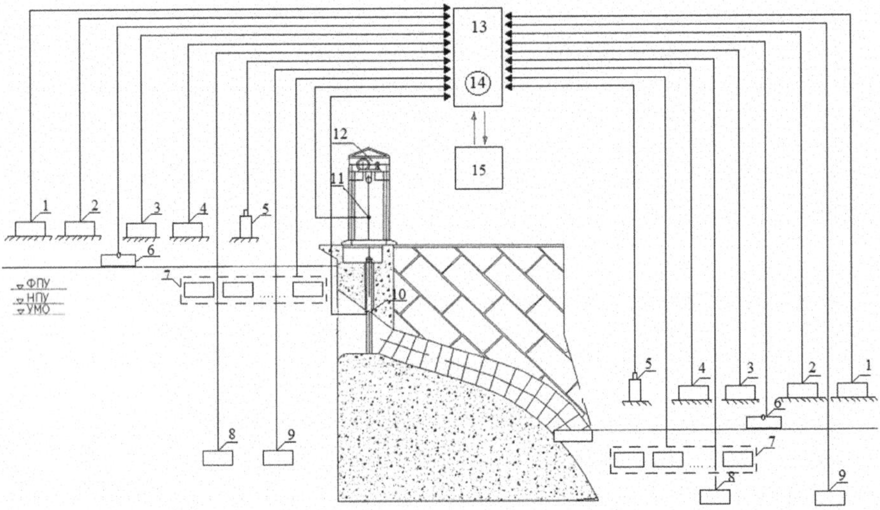

Fig. 1 System for automatic control of water level in a hydroelectric power station reservoir

loads in the power system recorded by the computer 13) with the actually recorded values of each of the two output quantities (water level in the reservoir and reservoir inflow flow). The arrays obtained in this way are referred to in the theory of neuroinformatics as "textbooks" (retraining implies constant updating of "textbooks" over time). Next, using standard algorithms (for example, an algorithm based on the back-propagation method), the neural processor 14 performs training, i.e. adjusting the weights of synaptic connections of the ANN, after which, upon reaching the permissible error (continuously comparing the output value reflected in the "textbook" with the value generated by the ANN), it stops training. After this, the neural processor 14 polls the trained ANN for future values of each of the two quantities (reservoir water level and reservoir inflow flow) based on the current input data, i.e. carries out forecasting with a given lead time of N days. As the actual values of these two forecast quantities are recorded (i.e., as we get closer to the forecast date), it becomes possible to compare the actual and ANN-generated values, as well as update the textbook and retrain the ANN.

With a forecast generated by the neural processor 14, according to which in N days the water level in the reservoir may exceed the maximum level and the computer 13 in this case controls the decrease in the water level, namely the computer 13 sends a command to the electric drive 12 of the spillway gate 10 to increase the flow of water. For each of the subsequent days before the predicted flood, the artificial neural network 14 also shifts the forecast window by one day and performs a new (refined) forecast of the above two values, adjusting the previous water release plan before the flood, as a result of which the computer 13 issues another command to the electric drive 12 gate 10. Thus, a smooth regulation of the water level in the reservoir before the flood is realized in order to avoid sudden discharges of water during the flood, and instead of directing the water past the turbines (idle discharges), the amount of generated electricity also smoothly increases.

During a flood, the reservoir's water level control system operates in enhanced mode. Units in the upper reservoir of a hydroelectric power station operate similarly to units in the lower reservoir. The proposed water level control system monitors the water level in the reservoir using sensors 1–9. The obtained data is processed by a computer 13, which uses the algorithm embedded in it to determine the instantaneous v_i (at the i-th moment of time) and the root-mean-square v_{av} rates of change in the water level in the reservoir. This algorithm consists of performing calculations using the formulas [2]

$$v_{cp} = \sqrt{\frac{v_i^2 + v_{i+1}^2 + \cdots + v_n^2}{n}},$$

$$v_i = \lim_{\Delta t_i \to 0} \frac{\Delta x_i}{\Delta t_i}, \quad (1)$$

where Δx_i is the change in water level in the reservoir over a period of time Δt_i; n—number of measurements.

Thus, the computer 13, equipped with a neural processor 14, performs data processing functions and calculates the rate of change of the water level in the reservoir and makes a decision on regulating the gate 10. For this, the value of the root-mean-square rate of change of the water level is used, which determines the speed of opening/closing of the gate. If the root mean square rate of change in water level is large, the time delay will be shorter and the speed of opening/closing the shutter will be higher. Along with this, using the speed value provides the ability to quickly control the shutter. If the water level exceeds the optimal highest mark of the lower threshold level (LTL) of the water surface of the reservoir, the system goes into high alert mode and checks the state of gate 10. After analyzing data on the root-mean-square rate of change in the water level in the reservoir, calculated from the indicators of sensors in the reservoir, the system determines the necessary shutter opening speed 10 and delay time to prevent accidental or too frequent operation of the control system. Following this, the system makes a decision to open the shutter 10 with the previously determined parameters. During regulation, the system continuously monitors parameters: water level at different points, the balance of water inflow and flow, and the serviceability of the elements of the regulation system. When the system establishes a decrease in the water level below the NPU, the possibility of closing the gate 10 is checked, in which the control system determines the presence or absence of flood, comparing data on the average rate of change in the water level, calculated from the readings of sensors in the reservoir. If the system detects a threat to coastal areas during a flood, it automatically opens the gate 10 to its maximum value to quickly release excess water from the reservoir. If the system determines that opening the valve 10 to its maximum value is necessary, but could lead to negative consequences, for example, the destruction of a dam or flooding of populated areas, then the system may decide to gradually open the valve 10 with a time delay to prevent catastrophic consequences. During operation, the system continuously monitors the position of the valve 10, the water level at different points and the serviceability of the elements of the control system. If any element of the control system malfunctions, the system automatically goes into emergency stop mode to prevent possible accidents and disasters. The interaction of the operator with the proposed system for the possibility of making adjustments to the technological regulation process, carrying out diagnostics of equipment and monitoring the quantities controlled by the system is carried out through a computer 13.

3 Setting Up an Information Measuring System

To set up the ANN, specific data from the Koka hydroelectric station in Ethiopia was used. In the process of selecting an ANN, models with different numbers of nodes and layers were considered, and a statistical analysis of the reliability of the models was carried out. The results of the analysis showed that the considered models satisfactorily predict the water level in the reservoir. When creating a water level forecasting model, information about hydrological conditions and water levels collected by a weather station over a certain period of time was used. A training set was used to train the neural network, and a control set was used for testing. The Python program [3] automatically normalized the data in accordance with the requirements for feeding it into the input of the neural network.

When designing a neural network, a formula was used that is a consequence of the Arnold–Kolmogorov–Hecht-Nielsen theorems, which determines the minimum number of neurons required to approximate any continuous function on a given compact set with a given accuracy. The formula is as follows [4]:

$$\frac{N_y \cdot Q}{1 + \log_2 Q} \leq N_w \leq N_y \left(\frac{Q}{N_x} + 1 \right) (N_x + N_y + 1) + N_y, \tag{2}$$

where N_y—is the dimension of the output signal; Q—number of elements of the training set; N_w—required number of synaptic weights; N_x—input signal dimension.

Using this formula, the following results were obtained: $3.76 \leq N_w \leq 37.5$.

In this case, the number of neurons in the hidden layer of a two-layer perceptron is determined by the formula:

$$N = \frac{N_w}{N_x + N_y} \tag{3}$$

Using the above formulas with $N_x = 3$, $N_y = 1$, $Q = 20$, the minimum and maximum values of $N = 1$ and $N = 9$ were obtained.

4 Conclusion

In this study, the technical result is achieved by the fact that the SEMS automatic water level control system in a hydroelectric power station reservoir contains sensors of physical quantities of all factors affecting the water balance, a control computer with a neural network, a spillway gate with an electric drive and a position sensor, which allows you to control the water level in the reservoir. The main difference between the developed system and existing ones is the use of an artificial intelligence system based on neural networks. Model results were verified against target values, confirming the validity of the models for water level forecasting. Therefore, the use of neural networks can be an effective approach for identification, control and prediction in various tasks. Thus, the development of a model for predicting the water level in the Koka reservoir using an artificial neural network is not only effective, but also a promising approach for solving such problems and can be applied in other areas of human activity.

References

1. Khaikin, S.: Neural Networks: A Complete Course, 2nd ed., 1104 p. (Kussul, N.N., ed.); lane from English Kussul, N.N., Shelestova, A.A. rev. M.: Williams (2006)
2. Kozyrev, A.V., Lashin, A.V., Semenov, V.A., Turunin, V.I.: System for automatic control of water level in a hydroelectric power station reservoir. RU 2629456, appl. 20 May 2016, publ. 29 Aug 2017
3. Ilyichev, V.Yu., Kachurin, A.V.: Creating Python programs to study the Mandelbrot set. E-Scio **5**(56), 362–371 (2021)
4. Lozovoy, Ya.S., Sekirin, A.I.: Solving the forecasting problem using neural networks [Electronic resource]. http://www.rusnauka.com/1_NTO_2011/Informatica/78176.doc.htm. Access date 20 Feb 2024

Formation and Analysis of Databases and Knowledge

Creation of Databases and Formalization of Knowledge Using Identification of SEMS, Working Under Various Constrains

Andrey Yu. Kuchmin

Abstract *Problem statement* The article is devoted to the development of methods for analysis of databases and formation of knowledge for constrained SEMS by identifying control systems of smart electromechanical systems. Of primary interest are developments in the field of real-time information gathering systems and control of complex physical and technical systems, which in the vast majority of cases are dynamic objects operating under conditions of all kinds of limitations. The article analyzes three directions in the identification of dynamic objects (DO): statistical, multi-functional and intellectual approaches. Modifications of identification methods for dynamic objects with constraints, formulated as problems of mathematical programming, are proposed. *Purpose* Development of identification methods for dynamic systems such as SEMS, which operate under various constraints. Reducing methods for identifying SEMS with models to a sequence of solving mathematical programming problems with simplified constraints. *Results* Methods for identifying dynamic control systems for complex objects taking into account various constraints based on a multiple-functional approach, which boil down to solving a sequence of mathematical programming problems. *Practical significance* The developed identification methods can be used in robotic systems created on the basis of SEMS modules.

Keywords SEMS · Control · Model · Dynamics · Identification · Constraint

---

A. Yu. Kuchmin (✉)
Institute for Problems in Mechanical Engineering of the Russian Academy of Sciences, Bolshoy Pr. V.O., 61, Saint-Petersburg 199178, Russia
e-mail: radiotelescope@yandex.ru

I. L. Tarasova and B. A. Kulik (eds.), *Smart Electromechanical Systems*, Studies in Systems, Decision and Control 544, https://doi.org/10.1007/978-3-031-64277-7_5

1 Introduction

Information-measuring and control systems (IMCS) in the modern world are widespread, and investments in this scientific and technical industry are increasing every year. In such systems, we see a convergence of information processes occurring in the IMCS and the behavior of objects of the physical world that are objects of control or monitoring. The joint description of information flows and physical processes in complex systems is possible only with cybernetic principles [1–3].

In 2006, Helen Gill proposed a new term—"cyber-physical systems" [4]. Cyber-physical systems integrate the ability to compute, transmit and store information with the monitoring and control of objects of the physical world. The main tasks in the field of theory and practice of cyber-physical systems are to ensure the efficiency, reliability and safety of functioning in real time.

Cyberphysical systems belong to a new scientific field—cybernetic physics, which, as defined by the author of this direction A. Fradkov in his book [2], is aimed at the study of physical systems by cybernetic methods.

The one rapidly developing direction of cyberphysical systems is smart electromechanical systems (SEMS) [5, 6]. The main objects of research are smart electromechanical systems. SEMS with actuators of parallel structure are considered.

SEMS began to be widely used in the 2000s in parallel robots or so-called parallel kinematic machines (PKM) [5–7], representing very good capabilities in terms of accuracy, rigidity and the ability to manipulate large loads. The basic component of the SEMS is the base unit, which is an adaptive platform moved by electromechanical jacks (actuators). Currently, SEMS are widely used not only in intelligent robots, but also in astronomy, machine tool building, medicine and other fields. The advantages, applications and descriptions of these mechanisms are reflected in publications [5–15].

Depending on the purpose of the SEMS modules, the type of actuators will also change. PKM in the form of an n-pod (a special case, this is the Stewart platform, or hexapod), which is an adaptive platform moved by linear propulsion. Most often, such a PKM consists of two rigid platforms connected by electrical actuators by means of hinges. This connection allows you to provide movement along six degrees of freedom. An alternative type of PKM has not rigid, but composite platforms that can change their size through electrical actuators.

The structure of the measuring system of the SEMS module can have different configurations depending on the type of PKM. To unambiguously determine the state of the PKM, it is necessary to use a laser measuring system of support bases on the lower and upper platforms. The second option provides for the presence of sensors for linear movements of electrical actuators and sensors for angular position in the hinges.

Models of varying degrees of complexity are used at different levels of decision-making in the hierarchical control system of the SEMS. Solving the problems of identifying such models should take into account the physical adequacy of the processes occurring in the SEMS, which can be done by introducing special constraints, for example, on the maximum and minimum values of the parameters of the SEMS model, the laws of conservation of momentum, energy, etc. It is important to take into account the ranges of values of the state vector of the object when identifying its parameters.

Identification methods are focused on working with digital information received from a digital measuring system. The methods are based on statistical identifiers based on the methods of least squares and multiple-functional methods, which take into account the peculiarities of the behavior of dynamic objects and changes in their parameters in the form of constraints described by systems of inequality. And as a target criterion for the quality of identification, extremes of some functionals are used. Such a formulation of the identification problem is convenient for implementation by methods of mathematical programming (MP).

2 Analysis of Methods for Forming of Database and Knowledge by Identifying of SEMS Control Systems

The general problems of identification of dynamic objects are considered. The following main factors hindering the effective application of classical approaches have been identified, such as non-stationarity and nonlinearity of DO, low information saturation of measuring information, the presence of non-stationary interference, delay in data transmission channels and digital processing devices. It is important to preserve the physical adequacy of the model to the processes taking place in DO and the feasibility of the laws of control.

Building a DO model is a process that boils down to forming the structure of the model with the subsequent adjustment of its parameters. The central problem of constructing models adequate to the real object is the identification of not only the parameters of the object, but also its structure, state and external influences, which in the general formulation is a nonlinear problem of large dimension [16–21]. Therefore, to simplify the task of identification, it is customary to divide it into a number of interrelated stages: the definition of the structure of the model, parameters, state and external influences.

In practice, the conversion of discrete samples with different frame rates is carried out using stacks with linear data interpolation. A mathematical description of the stack with linear interpolation of data in the form of a matrix shift operator in the time domain is proposed:

$$\mathbf{A}^{\tau} = \begin{cases} \mathbf{A}^{\mathrm{fix}\left(\frac{\tau\omega_0}{2\pi}\right)}\left[1 - \frac{\tau\omega_0}{2\pi} + \mathrm{fix}\left(\frac{\tau\omega_0}{2\pi}\right)\right] + \left(\frac{\tau\omega_0}{2\pi} - \mathrm{fix}\left(\frac{\tau\omega_0}{2\pi}\right)\right)\mathbf{A}^{\mathrm{fix}\left(\frac{\tau\omega_0}{2\pi}\right)+1}, & \tau \geq 0 \\ \left(\mathbf{A}^{\mathrm{T}}\right)^{\mathrm{fix}\left(-\frac{\tau\omega_0}{2\pi}\right)}\left[1 + \frac{\tau\omega_0}{2\pi} + \mathrm{fix}\left(\frac{-\tau\omega_0}{2\pi}\right)\right] + \left(\frac{\tau\omega_0}{2\pi} + \mathrm{fix}\left(\frac{-\tau\omega_0}{2\pi}\right)\right)\left(\mathbf{A}^{\mathrm{T}}\right)^{\mathrm{fix}\left(-\frac{\tau\omega_0}{2\pi}\right)+1}, & \tau < 0 \end{cases},$$

$$\mathbf{A} = \begin{bmatrix} 0\,0 & & 0 \\ 1\,0 & & 0 \\ & \ddots & \\ 0\,0 & 1 & 0 \end{bmatrix}_{n\times n}, \tag{1}$$

where τ is the shift in the time domain, ω_0 is the sampling frequency, fix(...) is the rounding function with the discard of the fractional part, n is the number of samples in the sample window. This operator is also applicable for the case of samples with a regular sampling rate. In this case, it will be:

$$\mathbf{A}^k = \begin{cases} \mathbf{A}^k, & k \geq 0 \\ \left(\mathbf{A}^{\mathrm{T}}\right)^k, & k < 0 \end{cases}, \quad \mathbf{A} = \begin{bmatrix} 0\,0 & & 0 \\ 1\,0 & & 0 \\ & \ddots & \\ 0\,0 & 1 & 0 \end{bmatrix}_{n\times n}, \tag{2}$$

where k is the number of samples by which the sample data is shifted can be both positive and negative. A negative value k means a delay, a positive value means ahead. This operator has a number of interesting properties, with the help of which, for example, it is convenient to calculate the finite differences of discrete samples or solve difference equations.

All existing approaches to identification can be divided into three groups: multiple-functional, statistical and intellectual [16–38]. This division is due to differences in the accounting of perturbations acting on the system.

Statistical methods include: correlation methods and methods of regression analysis, contrast method, Kalman filter, least squares identifiers, etc. [17, 20–22, 27–29, 34, 36].

Multiple-functional methods include: methods of identification by transient characteristics, deterministic identifiers, identifiers based on the velocity gradient method, identifiers based on integral transformations, methods of optimal identification, interval identification methods, etc. [16–19, 22, 30, 32, 33, 35–41].

Intelligent methods include: neural network methods and methods of hybrid networks (MANFIS, interval neural networks, etc.), methods of fuzzy logic, methods of evolutionary systems, logical-probabilistic methods, etc. [19, 23].

Methods from each group have their advantages and disadvantages. A widely used is a statistical approach based on the assessment of the parameters and state of the DO. The "flexibility" of the models is taken into account by the random nature of uncertain factors and interference, the models of which are given in the form of distribution laws, as a result, it is necessary to use large amounts of data with high information saturation. Most methods are based on various modifications of the method of least squares (MLS) [17] and require the use of powerful computing tools, the development of stable methods for estimating the probabilistic characteristics of perturbations.

Parametric identification algorithms based on the method of least squares (LS identifiers) find wide practical application. The LS identifier has the ability to flexibly change the volume of the regression sample and is not critical of the presence of noise in the measurements. LS identifiers are characterized by two types of information processing, according to one of which information is preliminarily accumulated (a sample is formed from a given number of input–output measurements), and then the problem of parametric estimation is solved by some suitable numerical method. Another type of LS identifiers involves the simultaneous updating of the regression sample and the solution of the parametric estimation problem, taking into account the solutions obtained in the previous steps. The procedure constructed in this way is recursion, and the identifier is a recurring.

In the case of small data samples, the sensitivity of LS identifiers to the choice of initial conditions of the DO model and poor convergence are manifested. Classical LS identifiers do not take into account the limitations associated with the modes of operation of the DO and the physical adequacy of the model to the processes in the DO.

In addition to LS identifiers, methods based on the Kalman filter have become widespread [42]. Initially, the Kalman filter was used to identify unmeasured components of the state vector of dynamical linear systems in the presence of noisy measurements. Further development of this approach made it possible to use the extended Kalman filter to identify both the state of the system and its parameters [43]. The disadvantages include the fact that errors in the filter are described by white noise, which does not exist in nature. It is possible to make a modified Kalman filter, in which the errors will be represented by color noise. The least favourable interference distributions are used as the main statistical model. Such a description is extreme and does not take into account the real conditions of operation of the DO, which reduces the quality of the identifiers on small samples. Such methods also do not take into account the limitations when solving the problem of identification.

An alternative approach is multiple-functional identification methods [16, 20, 36–38, 41]. Various methods of parametric and non-parametric identification of DO based on this approach have been developed [16].

The most common results in the creation of an approach for estimating the parameters of DO models using adaptive methods belong to Krassovsky [44, 45].

The method of speed gradient of Fradkov [39–41] became very widespread. This is a relatively easy-to-implement method, the effectiveness of which has been proven by its use in many applications. This method is unchallenged. The disadvantages of the method include unproven convergence in the presence of limitations and in the degeneration of the condition of reachability [41].

With the multiple-functional approach, accounting for constraints is possible if the identification problem is posed as a problem of the calculus of variations with constraints or as a problem of mathematical programming [46]. The idea of such methods is to determine a set of parameters or characteristics as a result of maximizing (minimizing) a certain functionality (often being a quadratic form between the response of a real system and the output of the model, the parameters of which are adjusted) by one of the known optimization methods, taking into account the corresponding constraints that specify the ranges of changes in the parameters of the system and other quantities, and are used to control the correctness of the model and its adequacy to nature. Such multiple-functional methods make it possible to describe the non-stationarity and stochasticity of DO by reducing to "non-rigid" models based on interval methods and systems of inequalities, including target ones. This approach can be used to jointly assess the structure, parameters and state of the DO. Such identification methods can be attributed to the category of optimization tasks that require complex calculations and specialized hardware for their implementation. Until recently, the application of such approaches in real tasks was constrained precisely for this reason.

Currently, many technical systems use neural networks for identification purposes. Neural networks were known in the 50s of the last century, but as an effective tool were not developed until the 80s, due to the lack of effective methods of their training and hardware implementation. Much attention is paid to the development of identification using neural and hybrid networks [19]. The central property of the neural network is its redundancy, which allows you to approximate various systems, this is the ability to predict their reactions [47]. There are two main problems in the synthesis of neural networks and their training—this is retraining and insufficient network dimensionality. Both of them are related to the issue of determining the dimensionality and topology of the network (the number of neurons, network layers, activation functions, etc.), training methods and data pre-processing at the design stage of neural networks. In practice, these problems are solved by the use of special heuristic techniques. This approach, due to the limited training sample at the design stage, can lead to network inoperability in real conditions.

In neural network technologies, identification is divided into three main areas—determining the topology of the network, training methods and data preprocessing (preprocessing). Often, the network topology is chosen for a specific task based on the accumulated experience of using this type of network in practice [19, 47–50]. Well-developed mathematical methods for selecting topologies of neural networks and comparing them have not been developed. The main practical approach is the use of combined (modular) networks, which consist of neural network subsystems of different topologies that solve the same problem, the results of which are accepted by voting. As teaching methods, the following are used: the method of reverse propagation of error, methods of optimization theory, methods of identification theory, adapted for a certain type of networks, evolutionary methods, etc. Preprocessing is a preliminary processing of data that allows you to highlight the main aspects of signals, remove noise and translate data into a basis convenient for further use in neural networks.

This work focuses on modifications of methods based on least squares [17, 21, 36], and multiple-functional methods [16, 20, 36].

It should be noted that the proposed identification methods have a high parallelization coefficient and can be implemented using neural network technologies, and limitations in the form of inequalities can be taken into account in the structure of interval neural networks. The author considers these technologies as further directions for the development of the proposed methods.

3 Formulation of Mathematical Programming Problems in the Identification of SEMS

For quasi-stationary SEMS, it is proposed to use identifiers with a model in the form of a stationary difference filter [36]:

$$\sum_{i_1=0}^{m_1} a_{i_1} x\left[t_k - \tau_{i_1}\right] = -\sum_{i_2=1}^{m_2}\sum_{i_3=0}^{m_{3,i_2}} b_{i_2,i_3} u_{i_2}\left[t_k - \tau_{i_2,i_3}\right], \tag{3}$$

where a_{i_1}, b_{i_2,i_3} are the filter coefficients, i_1, i_2, i_3 are the corresponding indices, m_1, m_2, m_{3,i_2} are the maximum values of the indices, t_k is the current time point, τ_{i_1}, τ_{i_2,i_3} are time shifts, x is the output coordinate, u_{i_2} are the control actions.

Using the matrix operator of shifting in the time domain, let's record the model in a compact form relative to the residual, convenient for the synthesis of identifiers:

$$\begin{gathered}\sum_{i_1=0}^{m_1} \mathbf{A}^{-\tau_{i_1}} \mathbf{x} a_{i_1} + \sum_{i_2=1}^{m_2}\sum_{i_3=0}^{m_{3,i_2}} \mathbf{A}^{-\tau_{i_2,i_3}} \mathbf{u}_{i_2} b_{i_2,i_3} = \boldsymbol{\varepsilon},\ \mathbf{Fp} = \boldsymbol{\varepsilon},\\ \left[a_0, a_1, \ldots, a_{m_1}, b_{1,0}, b_{1,1}, \ldots, b_{m_2,m_{3,m_2}}\right]^{\mathrm{T}} = \mathbf{p},\\ \left[\mathbf{A}^{-\tau_0}\mathbf{x}, \mathbf{A}^{-\tau_1}\mathbf{x}, \ldots, \mathbf{A}^{-\tau_{m_1}}\mathbf{x}, \mathbf{A}^{-\tau_{1,0}}\mathbf{u}_1, \mathbf{A}^{-\tau_{1,1}}\mathbf{u}_1, \ldots, \mathbf{A}^{-\tau_{m_2,m_{3,m_2}}}\mathbf{u}_{m_2}\right] = \mathbf{F},\end{gathered} \tag{4}$$

where $\mathbf{x}$, $\mathbf{u}_{i_2}$, $\boldsymbol{\varepsilon}$ are the corresponding vectors of the output coordinate samples, control actions and residuals, $\mathbf{p}$ is the vector of the estimated parameters, $\mathbf{F}$ is the matrix of parameters.

When describing an SEMS as a quasi-stationary DO, the limitations of its modes of operation and physical laws should be taken into account. These constraints will change as the DO moves from one region of the phase space to another. In a linear formulation, they have the form: where $\mathbf{Gp} \geq \mathbf{R}$, $\mathbf{G}$ and $\mathbf{R}$ are matrices of coefficients. Then the problem of identifying the parameters of the DO model in the form of a stationary difference filter can be formulated as an MP problem:

$$J = \min_{\mathbf{p}}\left\{\mathbf{p}^{\mathrm{T}}\mathbf{H}\mathbf{p}\right\} \text{ under restrictions } \mathbf{Gp} \geq \mathbf{R}. \tag{5}$$

In small samples of measurement data, the quality of identification is significantly affected by the values of the initial conditions of the DO. To take into account the initial conditions, it is proposed to introduce a matrix $\mathbf{S}$, composed of zeros and ones, then the matrix $\mathbf{H}$ will have the form: $\mathbf{H} = \mathbf{F}^{\mathrm{T}}\mathbf{S}^{\mathrm{T}}\mathbf{S}\mathbf{F}$. Standard LS identifiers do not take into account the initial conditions of DO, algorithms operate with large samples of data, most of which relate to steady-state conditions. When identifying IEMS, the stability of parameters remains only within small data samples, and the identifier works according to most of them are in transient modes, so some of the equations in (4) are incorrect and should be excluded when forming the matrix $\mathbf{H}$. This is achieved by zeroing the corresponding equations in (4). The estimation of the initial conditions and parameters of the DO model in the form of a stationary difference filter is possible using a multiple-functional approach, which will be discussed below.

If you select the value of the first coefficient in $\mathbf{p}$ equal to 1, you can use the substitution to move from (5) to the well-known MP problem: under constraints If the matrix $\mathbf{H}^*$ is positively defined, then this quadratic programming problem can be solved by standard methods, for example, the Wolfe method [46].

$$J = \min_{\mathbf{p}^*}\left\{0.5\mathbf{p}^{*\mathrm{T}}\mathbf{H}^*\mathbf{p}^* + \Phi\mathbf{p}^* + c\right\},\ \mathbf{G}^*\mathbf{p}^* \geq \mathbf{R}^*. \tag{6}$$

In the case of a low signal-to-noise ratio and low information saturation of the data, the matrix $\mathbf{H}^*$ becomes not positively defined, and it is advisable to proceed to solving the problem by the method of inequalities.

To synthesize the control laws of the SEMS, the methods of the state space (SS) are used. The transition from the description of the DO from the model in the form of a filter (3) to a view in the SS is carried out using standard structural transformations. The simplest way to obtain the equations of an object in the SS is to write in the form of a canonical form of observability, the matrices of which are formed from coefficients (3). The state vector of such a system in the SS can be reconstructed using a full-order state estimator.

In the presence of other measured quantities y_j in addition to the main coordinate, they are expressed in terms of the state variables of the system x_i as where $y_j[t_k] = \sum_{i=1}^{n} c_{j,i}x_i[t_k]$, $c_{j,i}$ are the coefficients of the matrix of the measured output in the SS. Passing to the sample vectors of the components of the state vector $\mathbf{x}_i$ and the measured coordinate $\mathbf{y}_j$ we get:

$$\begin{gathered}\mathbf{y}_j - \sum_{i=1}^{n} c_{j,i}\mathbf{x}_i = \boldsymbol{\varepsilon},\ \mathbf{F}\mathbf{p} + \mathbf{y}_j = \boldsymbol{\varepsilon},\\ \left[c_{j,1}, c_{j,2}, \ldots, c_{j,n}\right]^{\mathrm{T}} = \mathbf{p}, [-\mathbf{x}_1, \ldots, -\mathbf{x}_n] = \mathbf{F}.\end{gathered} \tag{7}$$

The coefficients $c_{j,i}$ can be found as minimization of the goal function $J = \min_{\mathbf{p}}\left\{0.5\mathbf{p}^{\mathrm{T}}\mathbf{F}^{\mathrm{T}}\mathbf{F}\mathbf{p} + \mathbf{y}_j^{\mathrm{T}}\mathbf{F}\mathbf{p} + 0.5\mathbf{y}_j^{\mathrm{T}}\mathbf{y}_j\right\}$, under interval constraints on the parameters $\mathbf{p}$. If the matrix $\mathbf{F}^{\mathrm{T}}\mathbf{F}$ is positively defined, then this is a quadratic programming problem, and it can be solved by standard methods, for example, the Wolfe method.

In the case of a low signal-to-noise ratio and low information saturation of the data, the matrix $\mathbf{F}^{\mathrm{T}}\mathbf{F}$ becomes not positively defined, and it is advisable to proceed to solving the problem by the method of inequalities [36].

For non-stationary SEMS, the use of identifiers with the model in the form of a non-stationary difference filter [36]:

$$\sum_{i_1=0}^{m_1} a_{i_1}[t_k]x\left[t_k - \tau_{i_1}\right] = -\sum_{i_2=1}^{m_2}\sum_{i_3=0}^{m_{3,i_2}} b_{i_2,i_3}[t_k]u_{i_2}\left[t_k - \tau_{i_2,i_3}\right], \tag{8}$$

wherein the non-stationary coefficients in (8) are defined as outputs of dynamical systems:

$$\sum_{j_1=0}^{m_4} \mu_{i_1,j_1} a_{i_1}\left[t_k - \tau_{j_1}\right] = 0, \sum_{j_2=0}^{m_5} \mu_{i_2,i_3,j_2} b_{i_2,i_3}\left[t_k - \tau_{j2}\right] = 0, \tag{9}$$

where j_1, j_2 are the corresponding indices, and m_4, m_5 are their maximum values, μ_{i_1,j_1}, μ_{i_2,i_3,j_2} are the coefficients of the corresponding filters, τ_{j_1}, τ_{j2} are time shifts. Then for the LS identifier we will have an expression for the goal function:

$$\Phi(\mathbf{p}) = \varphi(\mathbf{p})^{\mathrm{T}} \mathbf{H}_{\varphi} \varphi(\mathbf{p}), \tag{10}$$

where

$$\mathbf{H}_{\varphi} = \mathbf{F}_{\varphi}^{\mathrm{T}} \mathbf{F}_{\varphi},$$
$$\mathbf{F}_{\varphi} = diag\left[\mathbf{A}^{-\tau_0}\mathbf{x}; \mathbf{A}^{-\tau_1}\mathbf{x}; \ldots; \mathbf{A}^{-\tau_1}\mathbf{x}; \mathbf{A}^{-\tau_{1,0}}\mathbf{u}_1; \mathbf{A}^{-\tau_{1,1}}\mathbf{u}_1; \ldots; \mathbf{A}^{-\tau_{m_2,m_3,m_2}}\mathbf{u}_{m_2}\right],$$
$$\varphi = \left[\varphi_0(\mu_0, \mathbf{a}_0); \ldots; \varphi_{m_1}\left(\mu_{m_1}, \mathbf{a}_{m_1}\right); \varphi_{1,0}\left(\mu_{1,0}, \mathbf{b}_{1,0}\right);\right.$$
$$\left. \ldots, \varphi_{m_2,m_3,m_2}\left(\mu_{m_2,m_3,m_2}, \mathbf{b}_{m_2,m_3,m_2}\right)\right],$$

the components of which are polynomial functions from the elements of **p**, which are defined according to:

$$\sum_{i_1=0}^{m_1} \varphi_{i_1}\left(\boldsymbol{\mu}_{i_1}, \mathbf{a}_{i_1}, t_k\right) x\left[t_k - \tau_{i_1}\right] + \sum_{i_2=1}^{m_2} \sum_{i_3=0}^{m_{3.i_2}} \varphi_{i_2,i_3}\left(\boldsymbol{\mu}_{i_2,i_3}, \mathbf{b}_{i_2,i_3}, t_k\right) u_{i_2}\left[t_k - \tau_{i_2,i_3}\right] = \varepsilon[t_k]$$

and are the outputs of the models (9), and $\varepsilon[t_k]$—the current error, $\boldsymbol{\mu}_{i_1}$, $\boldsymbol{\mu}_{i_2,i_3}$—the estimated parameters of the filter, $\mathbf{a}_{i_1}$, $\mathbf{b}_{i_2,i_3}$—the initial values of the estimated parameters in (9). The parameters $\boldsymbol{\mu}_{i_1}$, $\boldsymbol{\mu}_{i_2,i_3}$, $\mathbf{a}_{i_1}$, $\mathbf{b}_{i_2,i_3}$ form the vector **p**.

Then the problem of identifying the parameters of the IEMS model in the form of a non-stationary difference filter can be formulated in the form of an MP problem:

$$J = \min_{\mathbf{p}}\left\{\varphi(\mathbf{p})^{\mathrm{T}} \mathbf{H}_{\varphi} \varphi(\mathbf{p})\right\} \text{ under constraints } \mathbf{G}(\mathbf{p})\mathbf{p} \geq \mathbf{R}. \tag{11}$$

The choice of solution method depends on the type of goal function and constraints. Various methods were used in the work, for example, the transition from conditional to unconditional optimization was applied with further solution by the following methods: Gauss–Newton method, Levenberg–Marquardt method, genetic algorithm, sequential quadratic programming method, complex method of Box, etc.

One practical approach to simplifying tasks (10) is to use model time scaling (i.e., introducing the frequency of quantization over time as a function of variable scale factors). With this approach, it is assumed that the model of the object describes a stationary DO, but the samples are taken at different intervals. Then the expressions (10) are converted to a form where the functions in **the vector** φ will depend on the variables of time shifts and the parameters of the stationary filter. For example, in the case of oscillatory systems, the scale coefficients can be calculated by changing the periods of oscillation at the time interval in which the samples are determined.

In addition to modifications of LS identifiers, the paper proposes identification methods based on the multi-functional approach [36]. Such multi-functional methods make it possible to describe the non-stationarity and stochasticity of DO by reducing to "non-rigid" models based on interval methods and systems of inequalities. As auxiliary tools when using this approach, various transformations are applied over measurement information, which increases noise immunity of multiple-functional methods.

Let's introduce the Φ function of assessing the quality of the model. Often, Φ can be a quadratic form of coordinates or parameters describing the deviations of the model from the measured DO data, Φ can also be expressed as an energy criterion, and for control tasks Φ can be a Lyapunov function. The DO model can be specified both explicitly and implicitly.

For many types of SEMS, nonlinear non-stationary models of the form can be used:

$$\dot{\mathbf{x}} = \mathbf{F}(\mathbf{p}, \mathbf{x}, \mathbf{u}, \mathbf{f}), \tag{12}$$

where $\mathbf{x}$ is the vector of the state DO, $\mathbf{p}$ is the parameters DO, $\mathbf{u}$ is the control action, $\mathbf{f}$ is the perturbation effects. Select where $\Phi = \int_{t_0}^{t_k} (\mathbf{Cx} - \mathbf{y}_0)^{\mathrm{T}}\mathbf{M}(\mathbf{Cx} - \mathbf{y}_0)dt$, $\mathbf{C}$ is the matrix of the measured output values of the system, $\mathbf{M}$ is the norming matrix, the coefficients of which are located in the range of change in the components of the vector of measured quantities $\mathbf{y}_0$. For the general case in (12) are unknown as a function $\mathbf{F}$, and part of the component $\mathbf{x}$ and arguments $\mathbf{p}$ and $\mathbf{f}$. One of the standard approaches is the approximation (12) of some network structure that organizes numerical decomposition into a certain series, but the question of choosing the topology of such a structure, the adequacy of such a description to physical laws remains open.

If the matrix function $\mathbf{F}$ is known, the problem of determining the unmeasured components of the state vector $\mathbf{x}$ and the unknown $\mathbf{p}$ and $\mathbf{f}$ is formulated as follows:

$$J = \min_{\mathbf{p},\mathbf{x},\mathbf{f}} \left\{ \int_{t_0}^{t_k} (\mathbf{Cx} - \mathbf{y}_0)^{\mathrm{T}}\mathbf{M}(\mathbf{Cx} - \mathbf{y}_0)dt \right\},$$

with constraints due to the equations of the DO model (12) and constraints describing the physical adequacy of the model where $\mathbf{p}_{\min} \le \mathbf{p} \le \mathbf{p}_{\max}$, $\mathbf{x}_{\min} \le \mathbf{x} \le \mathbf{x}_{\max}$, $\mathbf{f}_{\min} \le \mathbf{f} \le \mathbf{f}_{\max}$, $G_i(\mathbf{p}, \mathbf{f}, \mathbf{x}) \le 0$, $i = 1, \ldots, m$, $\mathbf{p}_{\min}$ and $\mathbf{p}_{\max}$ are the minimum and maximum values of the parameters of DO, $\mathbf{x}_{\min}$ and $\mathbf{x}_{\max}$ are the minimum and maximum values of the components of the state vector of DO, $\mathbf{f}_{\min}$ and $\mathbf{f}_{\max}$—minimum and maximum values of external influences acting on DO, G_i—functions of constraints.

For a numerical solution, the original problem can be transformed to the MP problem:

$$J* = \min_{\mathbf{p},\mathbf{x}[0],\mathbf{f}} \left\{ \sum_{k=0}^{N} (\mathbf{Cx}[k] - \mathbf{y}_0[k])^{\mathrm{T}} \mathbf{M} (\mathbf{Cx}[k] - \mathbf{y}_0[k]) \right\}, \tag{13}$$

With the limitations of the discrete model of DO in the form of: and constraints describing the physical adequacy of the model: where $\mathbf{x}[k+1] = \mathbf{F} * (\mathbf{p}[k], \mathbf{x}[k], \mathbf{u}[k], \mathbf{f}[k])$ $\mathbf{p}_{\min} \leq \mathbf{p} \leq \mathbf{p}_{\max}$, $\mathbf{x}_{\min} \leq \mathbf{x} \leq \mathbf{x}_{\max}$, $\mathbf{f}_{\min} \leq \mathbf{f} \leq \mathbf{f}_{\max}$, $G_i(\mathbf{p}, \mathbf{f}, \mathbf{x}) \leq 0$, $i = 1, \ldots, m$, k is the reference number for the time of the corresponding value or parameter of DO.

A method for solving MP (13) based on the introduction of basic functions that carry information about both the parameters of the system and its initial state and external influences is proposed. Basic functions are approximations of the solution of the system of differential equations that define the DO model, taking into account physical regularities and corresponding constraints on the parameters **p** and **f**, as well as phase coordinates **x**. The choice of types of basic functions and the determination of their parameters is carried out according to experimental data. Basic functions in the general case have the form: where $\mathbf{z}_i = \sum_{j=1}^{n} \varphi_j(\mathbf{p}_i, \mathbf{u}_i, \mathbf{v}_i, t)$, $\mathbf{v}_i = \sum_{j=1}^{n} \boldsymbol{\rho}_j(\boldsymbol{\mu}_i, t)$, $\mathbf{z}_i$ are basic functions that approximate $\mathbf{x}_i$ such that $\mathbf{x}_i = \mathbf{z}_i + \varepsilon$, where ε is the model error, $\varphi_j(\mathbf{p}_i, \mathbf{u}_i, \mathbf{v}_i, t)$ are the approximating functions of solutions (more often polynomial, or approximated by polynomials based on Weierstrass's theorem), $\mathbf{v}_i$ are basic functions that approximate $\mathbf{f}_i$ in such a way that $\mathbf{f}_i = \mathbf{v}_i + \varepsilon$, where $\boldsymbol{\rho}_j(\boldsymbol{\mu}_i, t)$ are the approximating functions of the model of external influences, $\boldsymbol{\mu}_i$ are the corresponding parameters. The representation **of** $\mathbf{x}_i = \mathbf{z}_i + \varepsilon$ is substituted in (13) and then the transformed problem of the MP is solved. Basic functions can be specified explicitly, for example, by some formula, implicitly as a numerical solution to the system (12) or as systems of inequalities.

The problem (13) leads to the solution of the optimization problem of high dimensionality and complexity, which requires the creation and adaptation of existing optimization procedures.

4 Comparison of LS Identifier and Identifier Based on the Basic Function

Consider the application of the method based on basic functions to identify the parameters and state of a quasi-stationary DO. As a model of DO we will use a stationary spread filter with a stable duty cycle of samples:

$$\sum_{i=0}^{n} a_i x[k-i] = -\sum_{j=1}^{n}\sum_{l=1}^{m} b_{l,j} u_l[k-j], \tag{14}$$

where a_i, $b_{l,j}$ are the filter parameters, k is the current discrete time point, the index in square brackets indicates the discrete time point, n is the filter order, m is the number of input influences, u_l is the input effects, x is the output value of the filter.

We will find a basic function containing information about both the filter parameters and the initial state. To do this, let's pass to the description of the model (14) in the samples:

$$\sum_{i=0}^{n} a_i \mathbf{A}^{-i}\mathbf{x} + \sum_{i=1}^{n} a_i \mathbf{A}^{-i}\mathbf{D}\mathbf{x}_0 = -\sum_{j=1}^{n}\sum_{l=1}^{m} b_{l,j}\mathbf{A}^{-i}\mathbf{u}_l,\ \mathbf{D} = \begin{bmatrix} \mathbf{0}_{h\text{-}n\times n} \\ \mathbf{1}_{n\times n} \end{bmatrix}, \tag{15}$$

where $\mathbf{A}$ is the matrix shift operator in the time domain, $\mathbf{x}$, $\mathbf{u}_l$ are the reference vectors of the filter output value and the input effects respectively, $\mathbf{x}_0$ are the initial filter conditions, $\mathbf{1}_{n\times n}$, $\mathbf{0}_{\text{h-}n\times n}$—unit and zero matrices of corresponding dimensions. Define an expression for the output value of the difference filter using (15):

$$\mathbf{x} = -\left[\sum_{i=0}^{n} a_i \mathbf{A}^{-i}\right]^{-1} \sum_{i=1}^{n} a_i \mathbf{A}^{-i}\mathbf{D}\mathbf{x}_0 - \left[\sum_{i=0}^{n} a_i \mathbf{A}^{-i}\right]^{-1} \sum_{j=1}^{n}\sum_{l=1}^{m} b_{l,j}\mathbf{A}^{-i}\mathbf{u}_l. \tag{16}$$

Let's select (16) as the base function for identifying the parameters of the spread filter, then we can introduce the goal function of the quality of identification in the form of: $\Phi = (\mathbf{x} - \hat{\mathbf{x}})^{\mathrm{T}}(\mathbf{x} - \hat{\mathbf{x}})$, where $\hat{\mathbf{x}}$ is the measured reference vector of the controlled output value of DO. The vector of parameters to be determined consists of the filter coefficients and the vector of the initial conditions: $\mathbf{p} = \left[a_0, \ldots a_n, b_{1,1}, \ldots, b_{m,n}, \mathbf{x}_0^{\mathrm{T}}\right]^{\mathrm{T}}$. It is necessary to take into account the limitations on the physical adequacy of the model (16) and the limitations on the initial conditions and filter coefficients $\mathbf{Gp} \geq \mathbf{R}$, then the MP problem will take the form:

$$J = \min_{\mathbf{p}}\{\Phi\}, \tag{17}$$

with constraints as Eqs. (16) and $\mathbf{Gp} \geq \mathbf{R}$. The solution of the problem (17) is much more complex than for LS identifiers (5) and is not reduced to the problem of quadratic programming. To solve the problem (17), the paper used batch implementations of conditional optimization methods in the MatLab environment, based on the Gauss–Newton method, the Levenberg–Marquardt method, the genetic algorithm, etc.

A model study of the effectiveness of the LS identifier and identifier based on the basic function was carried out. As an object, a second-order DO was modeled with a transfer function in the form of an oscillatory link. The input and output values of the DO, which were affected by additive interference with the normal law of distribution, were measured. The results of identification are presented in Table 1.

Table 1 Comparison of the efficiency of the LS identifier and the identifier based on the basis function (BFI)

%, noise	0	5	10	15	20	25	30	35	40	45
MSE %, a_i(LS identifier)	0	5.3	35.4	53.7	57. 1	65.7	58.3	65.3	68.5	78.7
MSE %, a_i(BFI)	0	1	1	3	3	3	3	10.3	28.2	38.2
MSE %, b_i(LS identifier)	0	27	32.6	76.3	96.2	98.6	99.2	102.4	122.1	134.5
MSE %, b_i(BFI)	0	3	7.4	11.3	12.4	12.7	18.3	38	58	92
MSE %, IC (BFI)	0	7	8.7	9.1	10.5	10.8	11.3	11.8	14.9	17.8

Where IC are the initial conditions

Studies of the proposed identifier based on the basic function showed that it works more efficiently on small samples in transient modes, has greater stability in the presence of noise in the measured values.

This method and the LS identifier were used to identify the parameters of the control system of the hexapod PI H-810.

5 Conclusion

The article analyzes the main problems arising in the identification of modern information-measuring and control systems of dynamic objects built on cybernetic principles. It is shown that in modern cyber-physical systems, the identification process is an integral part of the control loops. The influence of restrictions on the process of identification arising from the functioning of such systems, which should be taken into account in the synthesis of control laws, is described.

The main approaches in the identification of DO control systems have been analyzed. Modifications of quasi-stationary and non-stationary LS identifiers for DO control systems are proposed. A method for identifying DO based on a multiple-functional approach is proposed. It is shown that the synthesis of the proposed identifiers is conveniently represented in the form of mathematical programming problems.

Acknowledgements The work was carried out within the framework of the state assignment of the Ministry of Science and Higher Education of the Russian Federation for the Institute for Problems in Mechanical Engineering of the Russian Academy of Sciences.

References

1. Wiener, N.: Cybernetics: Or Control and Communication in the Animal and the Machine. Hermann & Cie, Paris; MIT Press, Cambridge, MA (1948). ISBN 978-0-262-73009-9; 2nd revised ed., 212 p. (1961)
2. Fradkov, A.L.: Cybernetic Physics: Principles and Examples, 207 p. Nauka, St. Petersburg (2003). (In Russian)
3. Kravets, A.G., Bolshakov, A.A., Shcherbakov, M. (Eds.): Cyber-Physical Systems: Advances in Design & Modelling, 347 p. Springer Nature Switzerland AG (2020)
4. Kupriyanovsky, V.P., Namiot, D.E., Sinyagov, S.A.: Cyber-physical systems as the basis of the digital economy. Int. J. Open Inf. Technol. **2** (2016). https://cyberleninka.ru/article/n/kiber-fizicheskie-sistemy-kak-osnova-tsifrovoy-ekonomiki. Date of access: 16 March 2023. (In Russian)
5. Gorodetskiy, A.E.: Smart Electromechanical Systems, 277 p. Springer International Publishing Switzerland (2016)
6. Gorodetskiy, A.E., Kurbanov, V.G.: Smart Electromechanical Systems: The Central Nervous System, 266 p. Springer International Publishing AG (2017)
7. Glazunov, V.A.: Mechanisms of Parallel Structure and Their Application: Robotic, Technological, Medical, Educational Systems, 1036 p. (academician R.F. Ganiev, ed.). Institute of Computer Research, Izhevsk (2018) (In Russian)
8. Artemenko, Yu.N., Agapov, V.A., Dubarenko, V.V., Kuchmin, A.Yu.: Co-operative Control of Subdish Actuators of Radio-Telescope IUS, 2012. N 4. P. 2–9. (In Russian)
9. Volkomorov, S.V., Kaganov, Y.T., Karpenko, A.P.: Modeling and Optimization of Some Parallel Mechanisms, 32 p. New Technologies, Moscow (2010). (In Russian)
10. Glazunov, V.A., Koliskor, A.Sh., Krainev, A.F.: Spatial Mechanisms of Parallel Structure, 96 p. Nauka (1991). (In Russian)
11. Zenkevich, S.L., Yushchenko, A.S.: Fundamentals of Control of Manipulative Robots, 480 p. MGTU (2004). (In Russian)
12. Merlet, J.P.: Parallel Robots (Solid Mechanics and Its Applications). Springer, Berlin (2004)
13. Kheilo, S.V., Glazunov, V.A., Palochkin, S.V.: Manipulative Mechanisms of Parallel Structure. Dynamic Analysis and Management, 86 p. MGUDT (2014). (In Russian)
14. Agapov, V.A., Gorodetskiy, A.E., Kuchmin, A.Yu., Selivanova, E.N.: Medical microrobot. Patent for invention RUS 2469752 20.05.2011. (In Russian)
15. Artemenko, Yu.N., Badu, E.I., Gorodetskiy, A.E., Dubarenko, V.V., Kuchmin, A.Yu., Tarasova, I.L.: Antenna radio telescope. Patent for invention RUS 2421765 09.02.2010. (In Russian)
16. Karabutov, N.N.: Structural Identification of Systems: Analysis of Dynamic Structures, 160 p. MGIU (2008). (In Russian)
17. Iserman, R., Munchof, M.: Identification of Dynamic Systems. Introduction and Identification of Dynamic Systems. An Introduction with Applications, 705 p. Springer (2011)
18. Mzyk, G.: Combined Parametric-Nonparametric Identification of Block-Oriented Systems, 238 p. Springer (2014)
19. Butalis, I., Zeodoridis, D., Kottas, T., Christodoulou, M.A.: System Identification and Adaptive Control. Theory and Applications of the Neurofuzzy and Fuzzy Cognitive Network Models, 313 p. Springer (2014)
20. Graupe, D.: Identification of Systems, 304 p. Krieger R. Publishing Company (1979)
21. Ljung, L.: System Identification Theory for User, 2nd edn. (1999)
22. Gusev, S.A.: Parametric Identification of Dynamic Objects. Methodical Instructions for the Performance of Laboratory Work, 24 p. SPB: SUAI (1997). (In Russian)
23. Zhumanov, I.I., Yusupov, N.R.: Optimization of identification of non-stationary objects on the basis of soft computing tools. Sci. World **2**(5(69)), 30–32 (2019). (In Russian)
24. Belov, M.P., Fyuong, Ch.Kh., Kozlova, L.P.: Method of identification of external disturbing effects on electric drives of a large radio telescope. Electrotechnica **12**, 44–49 (2019). (In Russian)

25. Prokopyev, I.V., Sofronova, E.A.: Investigation of the method of identification of the model and methods of controlling an unmanned vehicle by spatial trajectory. Reliab. Qual. Complex Syst. **3**(31), 99–111 (2020). (In Russian)
26. Misyurin, S.Y., Kreinin, G.V., Nosova, N.Y., Nelyubin, A.P.: Selection of a friction model to take into account the impact on the dynamics and positioning accuracy of drive systems. In: Advances in Intelligent Systems and Computing, vol. 1310, pp. 309–319 (2021)
27. Motorin, A.V., Koshaev, D.A., Stepanov, O.A., Yashnikova, O.M.: Identification of the model and smoothing of horizontal components of gravity perturbation for solving the problem of vector gravimetry. In: Materials of the XXIII Conference in Memory of the Outstanding Designer of Gyroscopic Devices, N. N. Ostryakov, St. Petersburg, pp. 149–152 (2020). (In Russian)
28. Basin, M.V., Stepanov, O.A., Litvinenko, Yu.A., Vasiliev, V.A., Tupysev, V.A.: Algorithm of polynomial filtration in the presence of nonlinearities in equations for measurements and state vectors. In: Materials of the XXXII Conference in Memory of the Outstanding Designer of Gyroscopic Devices, N. N. Ostryakov, St. Petersburg, pp. 316–321 (2020). (In Russian)
29. Vasiliev, V.A., Stepanov, O.A., Tupusev, V.A.: Analysis of the efficiency of the filtering algorithm of the discrete Markov process described by the forming filter of the polynomial type. In: XXVIII International Scientific and Technical Conference "Modern Technologies in the Problems of Control, Automation and Information Processing", pp. 167–168 (2019). (In Russian)
30. Dang, Kh.B., Pyrkin, A.A., Bobtsov, A.A., Vedyakov, A.A.: Identification of polynomial parameters of non-stationary linear systems. Proc. Higher Educ. Inst. Instrum. **64**(6), 459–468 (2021). (In Russian)
31. Bobtsov, A., Ortega, R., Nikolaev, N., Yi, B.: Adaptive state estimation of state-affine systems with unknown time-varying parameters. Int. J. Control (2021)
32. Korotina, M.M., Aranovsky, S.V., Bobtsov, A.A.: Evaluation of the parameters of the sinusoidal signal with the power function of frequency. Proc. Higher Educ. Inst. **63**(6), 507–514 (2021). (In Russian)
33. Le, V.T., Korotina, M.M., Bobtsov, A.A., Aranovsky, S.V., Vo, K.D.: Identification of linearly changing parameters in time of non-stationary systems. Mechatron. Autom. Control **20**(5), 259–265 (2019). (In Russian)
34. Jian, V., Bobtsov, A.A., Kolyubin, S.A., Pyrkin, A.A., Tuan, L.V.: Identification of piecewise linear parameters of regression models of non-stationary deterministic systems. Autom. Telemechanics **12**, 71–82 (2018). (In Russian)
35. Pervushin, V.F.: On nonparametric identification of linear dynamic objects. Bull. Tomsk State Univ. Control Comput. Eng. Inf. **4**(25), 95–104 (2013). (In Russian)
36. Kuchmin, A.Yu.: Identification of Non-Stationary Dynamic Objects. Informatsionno-upravliaiushchie sistemy [Information and Control Systems], no. 2, pp. 28–35 (2018). (In Russian). https://doi.org/10.15217/issn1684-8853.2018.2.28
37. Kuchmin, A.Yu., Dubarenko, V.V.: Definition of a rigidity of a hexapod. In: Gorodetskiy, A.E. (ed.) Smart Electromechanical Systems, pp. 159–163. Springer International Publishing Switzerland (2016)
38. Boykov, I.V., Krivulin, N.P.: Methods of identification of dynamic systems. Softw. Syst. Theory Appl. **5**(23), 79–96 (2014). (In Russian)
39. Fomin, V.N., Fradkov, A.L., Yakubovich, V.A.: Adaptive control of dynamic objects, 448 p. Nauka, Ch. ed. Physics and Mathematics of Literature (1981). (In Russian)
40. Fradkov, A.L.: Speed gradient scheme and its application in adaptive control problems. Avtomat. i Telemekh. **9**, 90–101 (1979); Autom. Remote Control **40**(9), 1333–1342 (1980). (In Russian)
41. Andrievsky, B.R., Fradkov, A.L.: Speed gradient method and its applications. Avtomat. Telemekh. **9**, 3–72 (2021). (In Russian)
42. Kalman, R.E.: A new approach to linear filtering and prediction problems. J. Basic Eng. **82**(1), 35–45 (1960)

43. Kopytin, A.V., Kopytina, E.A., Matveev, M.G.: Application of the extended Kalman filter to identify the parameters of a distributed dynamic system. Bull. Voronezh State Univ. Ser. Syst. Anal. Inf. Technol. Voronezh **3**, 44–50 (2018). (In Russian)
44. Krasovsky, A.A.: Optimal algorithms in identification problems with an adaptive model. Autom. Telemechanics **12**, 75–82 (1975). (In Russian)
45. Krasovsky, A.A., Bukov, V.N., Shendrik, V.S.: Universal Algorithms for Optimal Control of Continuous Processes, 272 p. Nauka (1977). (In Russian)
46. Tabak, D., Kuo, B.: Optimal Control by Mathematical Programming, 237 p. Prentice-Hall (1971)
47. Gorodetskiy, A.E., Tarasova, I.L.: Control and Neural Networks, 400 p. Publishing house of St. Petersburg State Polytechnic University (2005). (In Russian)
48. Aggarwal, C.C.: Neural Networks and Deep Learning: A Textbook, 520 p. Springer (2019)
49. Galushkin, A.I.: Neural Networks: Basic Theory, 496 p. Hotline-Telecom (2012). (In Russian)
50. Russell, S., Norvig, P.: Artificial Intelligence: A Modern Approach, 3rd edn. Artif. Intell. **175**, 935–937 (2011). https://doi.org/10.1016/j.artint.2011.01.005

New Methods for Recognizing and Resolving Contradictions in Knowledge

Boris A. Kulik and Alexander Ya Fridman

Abstract *Problem statement*: When analyzing and processing external information by SEMS modules, it may be necessary to consider knowledge that contains contradictions. Previously, the authors proposed methods for dealing with paradoxes by means of E-structures as well as some methods for resolving contradictions in knowledge bases by adding a new attribute. The problem is that those results do not explain the reasons for many other paradoxes and some contradictions in knowledge. *Purpose of research*: Continue the search and justification of new methods for resolving contradictions and explaining paradoxes. *Results*: A new model of polysyllogistics-like reasoning is proposed on the basis of the laws of algebra of sets, in which the shortcomings of E-structures are corrected. New methods for analyzing and resolving contradictions are developed. These include, in particular, the replacement of a universal proposition with a particular one in contradictory reasoning and the identification of binary relations for which the self-reference leads to a paradox. *Practical significance*: The results obtained make it possible to furnish the knowledge base of SEMS modules with new methods for recognizing and resolving contradictions when analyzing incoming information in the form of knowledge and reasoning.

Keywords Contradictions · Algebra of sets · Polysyllogistics · Law of paradox · Knowledge base · Relations · Self-reference

B. A. Kulik (✉)
Institute for Problems in Mechanical Engineering of the Russian Academy of Sciences, Bolshoy Pr. V.O., 61, 199178 Saint-Petersburg, Russia
e-mail: ba-kulik@yandex.ru

A. Y. Fridman
Institute for Informatics and Mathematical Modeling, Kola Science Centre of RAS, 24A Fersman Str., 184209 Apatity, Russia
e-mail: fridman@iimm.ru

I. L. Tarasova and B. A. Kulik (eds.), *Smart Electromechanical Systems*, Studies in Systems, Decision and Control 544, https://doi.org/10.1007/978-3-031-64277-7_6

1 Introduction

Analysis of the knowledge perceived by SEMS modules will be incomplete if their management systems do not provide methods for recognizing, analyzing and resolving contradictions. From the point of view of mathematical logic, reasoning that can be presented in the form of an identically false formula is considered contradictory one. For natural reasoning, as well as for many types of knowledge bases, this criterion is very inconvenient to use, that is why we propose below other simpler inconsistency criteria, which can be ultimately reduced to the standard inconsistency criterion adopted in logical calculus.

To solve the problem, a system of logical analysis is developed based on algebra of sets. The well-known book by Courant and Robbins [1] shows that the laws of algebra of sets, which are identical to the laws of classical logic, can be justified without axioms. In the book [2], this topic is covered in more detail. There, two systems of logical analysis are introduced, which use the laws of algebra of sets, namely *Euler structures* (*E*-structures) and *n-tuple algebra.*

To better understand the essence of the proposed methods of analysis, consider the difference between a contradiction and a paradox. It is generally accepted that a paradox is a situation when two mutually exclusive statements are true for the same object. For example, Russell paradox proves that the set of all non-self-reference (i.e., not being the elements of themselves) sets is self-reference and at the same time non-self-reference.

The following simple situation also leads to a paradox. Suppose, a reasoning reveals that an object A has a property B and does not have it at the same time. This can be expressed in the form of two premises expressing the *inclusion relation* ($\subseteq$) of sets: $A \subseteq B$ and $A \subseteq \overline{B}$. Applying the law of contraposition to the second premise, we obtain $B \subseteq \overline{A}$. Besides, applying the law of transitivity to $A \subseteq B$ and $B \subseteq \overline{A}$ leads to $A \subseteq \overline{A}$. In *E*-structures, the resulting formula is called a *paradox collision.* There are two possible solutions for this:

(1) a paradox collision testifies that the set A is equal to the empty set ($A = \emptyset$);
(2) if $A = \emptyset$ is unacceptable according to the conditions of the problem, then the reasoning is considered incorrect.

A paradox also arises when the properties of an object are not *contradictory* ones (e.g., like "birds" and "all animals except birds"), but they are *contrary* entities, when for two non-empty sets B and C the relation $C \neq \overline{B}$ is satisfied, but at the same time $C \subseteq \overline{B}$ (for instance, "birds" and "mammals"). Let us look how this happens. Assume, the premises $A \subseteq B$ and $A \subseteq C$ are given. Moreover, the properties B and C are contrary ones, which can be expressed by the relation $B \subseteq \overline{C}$. We use this relation as the 3rd premise. Then, according to the law of transitivity, $A \subseteq B$ and $B \subseteq \overline{C}$ yield $A \subseteq \overline{C}$. As a result, we again got the paradox collision $A \subseteq C$ and $A \subseteq \overline{C}$, and, accordingly, $A \subseteq \overline{A}$.

In mathematical logic, the following situation can be called a paradox. Let premises $F \supset G$ and $F \supset \neg G$ be given, where F and G are logical formulas, while $\supset$

and $\neg$ are logical connectives of implication and negation, respectively. According to the law of contraposition, $G \supset \neg F$ follows from the formula $F \supset \neg G$, while the formulas $F \supset G$ and $G \supset \neg F$ yield

$$F \supset \neg F = \neg F \vee \neg F = \neg F.$$

This means that for the given premises, the statement expressed by the formula F is false, but if, according to the conditions of the problem, F is true, then we must add the formula F to the premises. In this case, the obtained reasoning will correspond to an identically false formula.

A paradox arises also from the anomaly of contradiction in knowledge bases [3], in which the rules are expressed as the following structures:

$$r_m : B_1 \wedge B_2 \wedge \ldots \wedge B_n \rightarrow A,$$

where r_m is the name of a rule, while $B_1, B_2, \ldots, B_n, A$ are atoms, and the symbol $\rightarrow$ is interpreted as an implication.

Consider how an anomaly of contradiction is recognized. Let there be two rules:

$$r_D : B_1 \wedge B_2 \wedge \ldots \wedge B_n \rightarrow D;$$
$$r_F : B_1 \wedge B_2 \wedge \ldots \wedge B_n \rightarrow F.$$

The contradiction arises because the conditions (antecedents) of the rules coincide, while the consequents D and F are contraries (i.e., $D \supset \neg F$ is a true statement). One example of a contradiction anomaly is a case in medical diagnostics when two fundamentally different diseases are characterized by the same values of some symptoms, but at least one symptom is not specified, the values of which are different for these diseases.

A similar situation was also identified in the analysis of presupposition [4]. For instance, a presupposition of the utterance "John was late for the lecture" is the sentence "John was on his way to college". Moreover, the "paradoxicality" of the presupposition lies in the fact that it is true, regardless of whether John was late for the lecture or arrived on time. It was shown in [4] that the presupposition paradox can be eliminated by finding a parameter called the *assertion* (utterance) *switch*. For example, in the case of John, who was late for a lecture, this parameter could be the reason for his lateness (or its absence in the case of a timely arrival). Entering this parameter into the model in the form of a new logical variable leads to the elimination of the paradox. In contradiction anomalies for knowledge bases, the paradox can be eliminated by searching for a distinguishing parameter (e.g., by adding a symptom in the diagnostics of diseases, the meaning of which is different for different diseases).

To eliminate collision paradoxes in E-structures, two methods have been proposed [2]. The first is that in a sequence of inclusions (for example, $N_2 \subseteq N_3 \subseteq \overline{N_7} \subseteq N_5 \subseteq \overline{N_2}$) that leads to a paradox, we search for a proposition (let it be $N_2 \subseteq N_3$), which can be replaced by its inverse ($N_3 \subseteq N_2$) when, depending on the situation or our ignorance, the subject becomes a predicate or vice versa. There is an example in [2] where the premise "All brawlers are not self-confident" is replaced with "All not self-confident are brawlers". Different universes may have different correlations. After such a replacement, the paradox is eliminated.

The second method eliminates the inaccuracy in the formulation of a paradox. For instance, in one of the variants of the "Liar" paradox, the collision of the paradox is due to the fact that the proposition "All Cretans are not liars" is accepted as a negation of the statement "All Cretans are liars", although in fact the exact negation of the first statement is the proposition "Some Cretans are not liars". If this proposition is used as a negation, the paradox does not appear.

Note that the theoretical bases of E-structures are axioms with the help of which one of the types of partially ordered sets is formed. This article will consider an improved (compared to E-structures) version of the logical analysis systems, which is based on the laws of algebra of sets and is called the *mathematical model of polysyllogistics*. For this new version, new methods for analyzing and resolving contradictions will be proposed.

2 Mathematical Model of Polysyllogistics

We will call *polysyllogistics* a system of reasoning with an arbitrary number of premises in the form of Aristotelian propositions, unstructured relative to each other. Syllogistics [5, 6], created by Aristotle, is based on 4 types of propositions:

A: All P are Q, example: "All crocodiles are reptiles".

I: Some P are Q, example: "Some students are athletes".

E: All P are not Q, example: "All giraffes are not amphibians."

O: Some P are not Q, example: "Some of my colleagues don't like criticism."

A, ***I***, ***E***, ***O*** are generally accepted designations for these types of propositions, while types ***A*** and ***E*** are called *universal* propositions, and types ***I*** and ***O*** are called *particular* ones.

Below we list some of the shortcomings of syllogistics and the proposed ways to eliminate them.

*Disadvantage 1. A s*yllogism has only 2 premises and one conclusion. It is proposed to use the same method to analyze systems with any number of premises for which more than one consequence or conclusion is possible. Thus, we move on to the general method of analyzing reasoning with an arbitrary number of premises.

Disadvantage 2. In syllogistics, all basic terms (in the examples, P and Q) must not be equal to the empty set. This prohibition significantly limits the possibilities of analysis, since it does not allow checking the existence of the analyzed objects (from the point of view of mathematics, a set does not exist if, for some given premises, its equality to the empty set is proven). It is proposed to make the reasoning analysis system more flexible: do not generally prohibit the equality of terms to the empty set, but prohibit equality to the empty set for some terms, if this is included in the conditions of the problem under study.

Disadvantage 3. The basic version of syllogistics does not allow a negation term to be placed at the beginning of a proposition (for example, the propositions "Some not-*P* are *Q*" and "All not-*P* are *Q*" are considered incorrect). The result of this prohibition is that, in syllogistics, some correct reasoning can be recognized incorrect ones. In the proposed system, this prohibition makes no sense.

Disadvantage 4. In syllogistics, when changing the order of premises, some incorrect modes of the syllogism are often formed, due to which such changing is prohibited. In the proposed system, when the order of the premises changes, the result of the logical analysis remains unchanged.

Disadvantage 5. The rules of syllogistics have no strict justification. In the proposed reasoning model, all analysis methods are based on the laws of algebra of sets.

Let us consider how a logical inference system can be built based on the laws of algebra of sets, taking into account the proposed corrections. Aristotelian types of propositions can be expressed by correlations of algebra of sets. Let the terms *P* and *Q* in the above examples of propositions correspond to some sets of the same name. Then the following expressions of algebra of sets will correspond to the types of propositions:

$$\boldsymbol{A} : P \subseteq Q; \boldsymbol{I} : P \cap Q \neq \emptyset; \boldsymbol{E} : P \subseteq \overline{Q}; \boldsymbol{O} : P \cap \overline{Q} \neq \emptyset.$$

Considering this, the following mathematical model is proposed for polysyllogistics. In a universe $\boldsymbol{U}$, let a system of sets S_i ($i = 1, 2, \ldots, n$) and their complements be given, and some of these sets can be equal to the empty set or the universe. It is also possible that as a result of the analysis, some of these sets will turn out to be equal to each other. For this system, two types of *logical premises* are specified:

Type 1*. Universal propositions* are specified as inclusion relations between some of the given sets (e.g., $\overline{S_2} \subseteq S_3$) and.

Type 2*. Particular propositions*, which express the truth of an non-empty intersection of some pairs of sets (for example, $S_2 \cap S_6 \neq \emptyset$).

This system also allows to set two types of restrictions:

- *Equivocation constraints* are expressed as the inadmissibility of equality for some pairs of sets (for instance, $S_2 \neq S_4$).
- *Emptiness constraints* are expressed as the inadmissibility of equality to empty sets for some of the given sets (say, $S_4 \neq \emptyset$).

Being defined this way, the reasoning model admits solving the following problems:

Task 1. Find consequences in the form of inclusion relations between some pairs of sets not specified in the conditions of the problem (thereby deducing consequences as new universal propositions).

Task 2. Check whether any equivocation constraints are violated in this system.

Task 3. Check whether any restrictions of emptiness are violated in this system (the absence or presence of paradoxes in the reasoning is checked).

Task 4. Find new pairs of sets for which their non-empty intersection is proven (thereby deducing conclusions in the form of new particular propositions).

In syllogistics, only Tasks 1 and 4 are solved with great difficulty and not without errors. Solutions to Tasks 1 – 4 based on E-structures can be found in the book [2]. The mathematical model proposed here differs from the one presented in this book, since it has no axioms, and all methods and rules are based on the laws of algebra of sets.

To solve these problems, the following well-known laws of algebra of sets are used as inference rules:

law of contraposition: $X \subseteq Y$ is equivalent to $\overline{Y} \subseteq \overline{X}$;

double complement law: $\overline{\overline{X}} = X$;

law of transitivity: if $X \subseteq Y$ and $Y \subseteq Z$, then $X \subseteq Z$;

law of equivalence: if $X \subseteq Y$ and $Y \subseteq X$, then $X = Y$.

However, to solve Tasks 3 and 4, the well-known 26 laws of algebra of sets, which are given in the book [1], turned out to be insufficient. The next section provides the formulation and rationale for two new laws of algebra of sets.

It is convenient to start solving all polysyllogistics problems by constructing an inclusion graph. Then their solution becomes significantly simpler.

A *literal* is a designation for a set or its complement, and it is clear that, for example, S2 and are different literals.

An *inclusion graph* is a drawing in which literals are represented by points (they are called *vertices* of the graph), while inclusion relations between sets are depicted as arrow lines called *arcs* of the graph. If, for example, $S_4 \subseteq \overline{S_2}$, then the inclusion graph contains the arc directed from the literal S_4 to the literal $\overline{S_2}$ ($S_4 \rightarrow \overline{S_2}$).

To simplify drawings, we will not depict points in the figures, as is customary in graph theory; we will just draw arcs from a literal to another literal.

The procedure for solving all 4 problems begins the same:

(1) draw an inclusion graph for all premises (how to depict particular propositions on the graph will be shown later when solving Example 2);
(2) apply the law of contraposition to all premises and complete the inclusion graph.

The resulting inclusion graph can now be used to analyze and solve all the polysyllogistics problems listed above.

Example 1 This example, slightly modified by us, is taken from the book [7]. In the proposed formulation, the term "pawnbroker" is replaced by the term "cheater" and a new premise is added. This example is interesting because it demonstrates a situation where an absurd consequence is not caused by errors in the logical inference, but is imperceptibly present in the premises given as follows:

(1) promise-breakers are not trustworthy;
(2) all wine-drinkers are very communicative;
(3) a person who keeps their promises is honest;
(4) no teetotalers are cheaters;
(5) one can always trust a very communicative person;
(6) all honest persons are not cheaters.

It is necessary to derive new universal propositions and verify the existence of fraudsters, i.e., check whether the emptiness constraint for the term "cheaters" is met.

Let us formulate the premises of the polysyllogism in the form of relations between the following sets: $\boldsymbol{U}$—people; S_1—promise-breakers; S_2—trustworthy persons; S_3—wine-drinkers; S_4—very communicative persons; S_5—honest persons; S_6 are cheaters.

Then the statements of Example 1 can be expressed in the form of the following relations.

$$(1)S_1 \subseteq \overline{S}_2;\ (2)S_3 \subseteq S_4;\ (3)\overline{S}_1 \subseteq S_5;\ (4)\overline{S}_3 \subseteq \overline{S}_6;\ (5)S_4 \subseteq S_2;\ (6)S_5 \subseteq \overline{S}_6.$$

Emptiness constraint: $S_6 \neq \emptyset$.

Now let us build the inclusion graph (Fig. 1).

It is better to depict the inclusion graph in this way: let all positive literals be located in the top line, while all negative literals are in the bottom line. Then we will not miss a single literal, and besides, on such a diagram, in which the opposite literals are located vertically, it is easy to deduce the contrapositions of the initial propositions.

Let us apply the law of contraposition (and, where necessary, the double complement law) to the initial premises. Then we get:

$$(7)S_2 \subseteq \overline{S}_1;\ (8)\overline{S}_4 \subseteq \overline{S}_3;\ (9)\overline{S}_5 \subseteq S_1;\ (10)S_6 \subseteq S_3;\ (11)\overline{S}_2 \subseteq \overline{S}_4;\ (12)S_6 \subseteq \overline{S}_5.$$

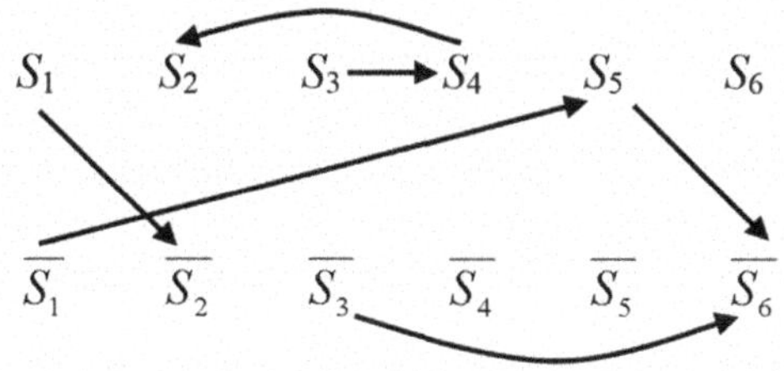

Fig. 1 Inclusion graph for the premises of Example 1

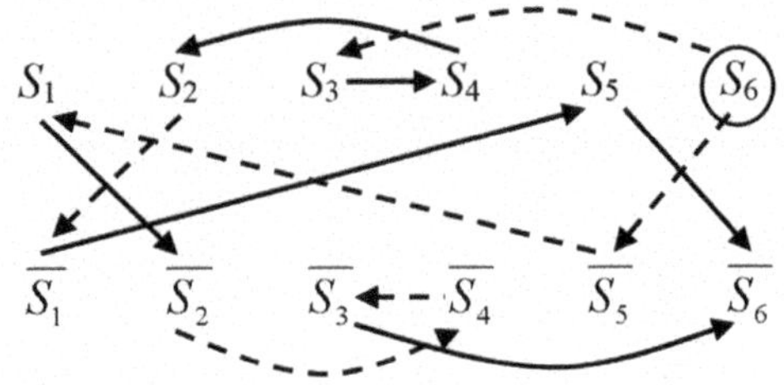

Fig. 2 Premises and consequences of Example 1

Construct the inclusion graph for the premises and their contrapositions (depicted by dashed lines in Fig. 2).

The figure shows that the contrapositions of the initial propositions can be easily constructed by using one simple rule: *if an initial proposition connects a pair of literals, then the contraposition of this proposition connects the opposite literals, and the direction of the arc (to the right or left) is reversed.*

Now it is easy to find a solution to the problem. In the inclusion graph, let us select literals having no input arcs (*initial literals*, in this example there is only one such literal S_6) and "walk" from them in the direction of the arcs. As a result, we get two *paths* in the graph (the second path is the contraposition of the first one):

$$S_6 \to S_3 \to S_4 \to S_2 \to \overline{S}_1 \to S_5 \to \overline{S}_6; \quad S_6 \to \overline{S}_5 \to S_1 \to \overline{S}_2 \to \overline{S}_4 \to \overline{S}_3 \to \overline{S}_6$$

The law of transitivity allows to state that if there is a path from one literal to another, then the set corresponding to the initial literal is included in the set corresponding to the final literal. From here, it is clear that in both cases we get $S_6 \subseteq \overline{S}_6$ that is the ***paradox collision***.

Moreover, if we replace the designations of literals with the corresponding terms, then a number of mutually exclusive statements will appear: cheaters are both teetotalers and drinkers, communicative and non-communicative persons, honest and dishonest ones, etc. Let us see whether the given constraint $S_6 \neq \emptyset$ is observed. In E-structures [2], the expression $S_6 \subseteq \overline{S}_6$ is equivalent to the fact that $S_6 = \emptyset$. This suggests that the emptiness constraint does not hold in this example.

However, the conclusion that a paradox collision implies the equality of the corresponding literal is to the empty set was not sufficiently substantiated in E-structures. This relation is not found in the known laws of algebra of sets [1]. Therefore, it is advisable to formulate and justify this law. At the same time, in the next section, another new law of algebra of sets will be introduced and proved, which is necessary in the construction and analysis of particular propositions.

It is worth noting that if in Example 1 we exclude the 6th proposition "all honest persons are not cheaters", then the paradox will not appear, but at the same time we will get the unexpected consequence "all cheaters are honest", which clearly does not correspond to the meaning of the word "cheater".

3 New Laws of Set Algebra

The need for new laws is caused by the fact that without them it is impossible to solve some problems of polysyllogistics. If we take into account that in the structures of n-tuple algebra, logical formulas correspond to sets of elementary tuples (i.e., essentially, sets of their satisfying substitutions), and the inclusion relation of sets is isomorphic to the relation of logical implication [2], then these laws are also applicable in logical inference problems in mathematical logic. The first law is called the law of paradox.

(i) ***Law of paradox***: if it is proven that $X \subseteq \overline{X}$, then $X = \emptyset$.

Proof From the definition of the complement operation it follows that the set $\overline{X}$ contains those and only those elements of the universe $\boldsymbol{U}$ that are not elements of X. Suppose that there exists an element a such that $a \in X$. At the same time, from the condition $X \subseteq \overline{X}$ it follows that the element a is contained in $\overline{X}$ as well, i.e., by definition, $a \notin X$. The resulting contradiction can be resolved in the only way: $X = \emptyset$. *End of proof.*

The second law is used in the cases where particular propositions are present or derived in the reasoning. Since the relation of inclusion for sets underlies logical inference, expressions for particular propositions of the type $S_2 \cap S_6 \neq \emptyset$ do not make it possible to obtain new particular propositions since it does not contain an inclusion relation, which prevents the usage of the transitivity law. Therefore, it was proposed to give the following formulation for particular propositions: the relation $X \cap Y \neq \emptyset$ is equivalent to two relations $\alpha_k \subseteq X$ and $\alpha_k \subseteq Y$, provided that $\alpha_k \neq \emptyset$. To justify this replacement, it was necessary to formulate another new law of set algebra.

(ii) ***Condition of non-empty intersection***: provided that $\alpha_k \neq \emptyset$, $\alpha_k \subseteq X$ and $\alpha_k \subseteq Y$ are equivalent to the expression $X \cap Y \neq \emptyset$.

Proof If $\alpha_k \neq \emptyset$, then there is an element b such that $b \in \alpha_k$. From $\alpha_k \subseteq X$ and $\alpha_k \subseteq Y$ it follows that $b \in X$ and $b \in Y$, hence, $X \cap Y \neq \emptyset$. *End of proof.*

4 Resolution of Contradictions by Converting a Universal Proposition into a Particular One

Let us try to reveal the reason for the absurdity in Example 1. The logical inference does not contain errors, so it is clear that the contradiction is implicitly contained in the seemingly plausible premises. Suppose that some of these premises, expressed as universal propositions, will be closer to the truth if reformulated as particular propositions. Accordingly, look at an example.

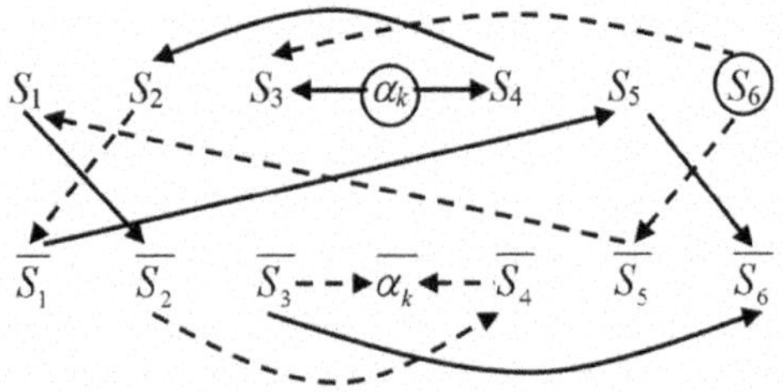

Fig. 3 Inclusion graph of Example 2

Example 2 In Example 1, we make the following change: we transform the second premise "all wine-drinkers are very communicative" into a more plausible particular proposition "some wine-drinkers are very communicative". Then, instead of the premise $S_3 \subseteq S_4$, two premises must be entered: $\alpha_k \subseteq S_3$ and $\alpha_k \subseteq S_4$. In accordance with this, we will construct the inclusion graph (Fig. 3).

Analysis of this inclusion graph allows to conclude the following.

(1) The inclusion graph has two initial literals: S_6 and α_k.
(2) Two branches come out from the literal S_6 (cheaters): $S_6 \rightarrow \overline{S_5} \rightarrow S_1 \rightarrow \overline{S_2} \rightarrow \overline{S_4} \rightarrow \overline{\alpha_k}$ and $S_6 \rightarrow S_3$. This means that cheaters have the following qualities: not honest, breaking promises, untrustworthy, uncommunicative, wine-drinkers. Among the listed properties, unlike Example 1, there are no contradictory ones, which mean the absence of a paradox.
(3) Two branches come out from the non-empty literal α_k: $\alpha_k \rightarrow S_4 \rightarrow S_2 \rightarrow \overline{S_1} \rightarrow S_5 \rightarrow \overline{S_6}$ and $\alpha_k \rightarrow S_3$. According to the condition of non-empty intersection, it turns out that pairs of literals located on one of these branches have a non-empty intersection (e.g., $S_4 \cap \overline{S_1} \neq \emptyset$, since $\alpha_k \subseteq S_4$ and $\alpha_k \subseteq \overline{S_1}$), as well as pairs of literals contained in different branches coming out from α_k (for instance, $\overline{S_6} \cap S_5 \neq \emptyset$, since $\alpha_k \subseteq \overline{S_6}$ and $\alpha_k \subseteq S_5$). It is easy to verify that among these pairs there is no pair (S_5, S_6), from which it follows that the particular proposition "some cheaters are honest" ($S_5 \cap S_6 \neq \emptyset$) is not a conclusion in this example.

Thus, replacing at least one universal proposition in the chain of inclusions, which leads to a paradox, with a particular proposition makes it possible to completely eliminate the logical catastrophe. It also follows that in a number of cases such a seemingly inconspicuous logical mistake as the transformation of particular propositions into universal ones leads to a paradox.

The method for resolving contradictions proposed in Example 2 can also be justified for the general case, which is expressed in the following theorem.

Theorem (elimination of a paradox by transforming a universal proposition into a particular one). If a contradiction in a reasoning is caused by a chain of inclusions of literals $L_1 \subseteq L_2 \subseteq \ldots \subseteq L_n$, under the condition $L_n = \overline{L_1}$, then the paradox can be eliminated by choosing a pair of literals (L_i, L_{i+1}) in this chain and replacing the relation $L_i \subseteq L_{i+1}$ with two relations $\alpha_k \subseteq L_i$ and $\alpha_k \subseteq L_{i+1}$.

Proof With the replacement provided for in the Theorem, the path leading from L_1 to $\overline{L_1}$ is broken in the inclusion graph, since in the initial chain the $L_i \rightarrow L_{i+1}$ is transformed into the subgraph $\overline{L_{i+1}} \rightarrow \overline{\alpha_k} \leftarrow \overline{L_i}$. In the chain, which is formed by the contrapositions of the arcs included in the original chain, and, like the original one, it represents the path from L_1 to $\overline{L_1}$, a break also occurs, since the arc $\overline{L_{i+1}} \rightarrow \overline{L_i}$ is replaced by the subgraph $\overline{L_{i+1}} \rightarrow \overline{\alpha_k} \leftarrow \overline{L_i}$, which prevents the literal $\overline{L_1}$ from being reached from the literal L_1. This proves that replacing the universal proposition $L_i \subseteq L_{i+1}$ with the particular proposition $L_i \cap L_{i+1} \neq \emptyset$ results in eliminating the paradox is. *End of proof.*

5 Analysis of Properties of Binary Relations in a Contradictory Reasoning

Let us consider some paradoxes of set theory. We start with a popular version of the main paradox of set theory, namely Russell paradox. A regimental barber was given orders to shave those and only those who did not shave themselves. The barber calmly shaved his fellow soldiers, without feeling a catch, but when it was time to shave himself, the illogicality of the order was fully revealed. If he shaves himself, it turns out that he is shaving someone who is shaving himself and thereby violates the order. If he does not shave, he will again violate the order, since he thereby refuses to shave someone who does not shave himself.

To understand the reason for this paradox, consider some aspects of the theory of binary relations. According to this theory, the relations can be *reflexive* and *anti-reflexive* ones. For instance, the relations "included or equal to" ($\subseteq$) and "not greater than" ($\leq$) are reflexive, since the correlations $A \subseteq A$ and $A \leq A$ are not in doubt. At the same time, for other relations, such as "strictly included" ($\subset$) and "less than" ($<$), reflection leads to a paradox (e.g., it is not clear how something can be greater than itself).

Let us consider the relation "shave" from this point of view. Is it possible to believe that the relation "shave himself" is a special case of this relation? A simple analysis shows that this is not true, and not only because shaving another person requires certain knowledge and skills. The fact is that "shave" and "shave himself" do not coincide in meaning. If a barber is shaving another person (suppose, a baker), then a situation is possible where at the same time the barber is shaving the baker, and the baker is shaving himself. Of course, no self-respecting barber would allow such outrage, but in principle such a situation is possible. Conversely, if a person shaves, then he cannot shave and not shave at the same time. This means that the "shaving" relation is anti-reflexive.

Based on this example, it can be assumed that many arguments and proofs that use reflexivity (self-referentiality in Russell's phrase) are based on relations for which reflexivity is refuted by detailed analysis. In other words, in some self-referential reasoning, the contradiction may be hidden in the source data.

Let us consider an informal formulation of Russell paradox. Let us call a set *self-applicable* one if it is its own element. The "set of all sets" is often cited as an example in publications. At the same time, *ordinary* sets are not self-applicable, for instance, a set of programmers is not a programmer.

Consider the set A_R of all ordinary sets (it is called the *Russell set*). Is it ordinary one?

There is no clear answer to this question. If the set A_R is an ordinary one, then it must be an element of the set A_R, thus making it self-reference. If the A_R is self-reference, then it should not be an element of the set A_R by the definition of this set. However, if it does not include itself as an element, then it is an ordinary set.

From the point of view of algebra of sets, the following explanation of this paradox is possible. First, in algebra of sets, the terms "set" and "element" mean significantly different concepts. They are linked by the membership relation ($a \in A$) rather than by the inclusion relation, which is used only for sets ($a \subseteq A$ is incorrect, the correct one would be $\{a\} \subseteq A$). Moreover, in algebra of sets there is no need to use sets as elements of other sets. The system-forming relation in algebra of sets is not the membership relation ($a \in A$), as in set theory, but the inclusion relation ($\subseteq$), for which the "self-reference" ($A \subseteq A$) does not lead to a paradox. In algebra of sets, the membership relation is used only to define certain sets and also to notify that a certain element belongs (or does not belong) to a certain set.

Secondly, if we nevertheless allow in algebra of sets the possibility of considering a set an element of another set and introduce "self-reference" for the membership relation, then according to the law of paradox, the Russell set A_R, by virtue of the previously proven law of paradox, will be equal to the empty set.

6 Conclusion

(1) A new polysyllogistics-like model of reasoning has been developed, which is based on the laws of algebra of sets. This model allows analyzing an unlimited number of premises corresponding to universal and particular propositions of Aristotelian syllogistics. Above this, the model can use *equivocation constraint* (prohibition of equality for some pairs of selected sets) and *emptiness restrictions* (prohibition of equality of selected sets to an empty set). The proposed model significantly expands the analytical capabilities of modeling and analysis of reasoning.

(2) New laws of algebra of sets are formulated and proven: the *law of paradox*, which justifies the equality of objects with contradictory or contradictory properties to an empty set, and the *condition for non-empty intersection of sets*, which allows expressing a statement about the non-empty intersection of sets by using the inclusion relation.

(3) New methods for resolving contradictions are proposed and justified, namely, transforming one of the premises from a universal proposition into a particular one and analyzing the reflexivity property for a relation present in the contradictory reasoning.

Acknowledgements The research was supported by the Ministry of Science and Higher Education of the Russian Federation (project no. 124041500008-1).

This work was carried out within the framework of the state task for the IIMM KSC RAS (R&D "Development of Theoretical, Organisational and Technical Foundations of Information Support for Management of Viability in Regional Critical Infrastructures in the Arctic Zone of the Russian Federation" No. FMEZ-2022-0023).

References

1. Courant, R., Robbins, H.: What is Mathematics? An Elementary Approach to Ideas and Methods (2nd ed.). Oxford University Press, p. 566, New York (1996)
2. Kulik, B., Fridman, A.: Complicated Methods of Logical Analysis Based on Simple Mathematics, p. 195. Cambridge Scholars Publishing, Newcastle upon Tyne (2022)
3. Harmelen, F.: Applying rule-base anomalies to KADS inference structures. Decis. Support Syst. **21**(4), (1997)
4. Kulik, B., Fridman, A.: Roles contradictions play in logical models of metaphors and presuppositions (2018) CEUR Workshop Proceedings 2303
5. Smith R.: Aristotle's logic, the Stanford encyclopedia of philosophy (Winter 2022 Edition), In: Zalta, E.N., Nodelman, U. (eds.) https://plato.stanford.edu/archives/win2022/entries/aristotle-logic/
6. Copi, I.M., Cohen, C., McMahon, K.: Introduction to Logic, p. 654. Routledge, New York (2016)
7. Carroll, L.: Symbolic logic. https://www.gutenberg.org/files/28696/28696-h/28696-h.htm

Intellectual Model of Mobile Robots Coalitions Formation in Network-Centric Group Control Systems

Vladimir A. Serov, Evgeny M. Voronov, and Yulia Yu. Kurilenko

Abstract *Problem statement* An important feature of the network-centric intelligent mobile robots group control system for is the requirement of structural-goal adaptation, which should provide the ability to solve a wide class of tasks under uncertain environment. One of the basic mechanisms of structural-goal adaptation is the formation of coalitions of mobile robots with a heterogeneous structure and focused on solving complex tasks corresponding to the current situation. The heterogeneity property of the mobile robots coalition means that the members of the coalition, from one side, have a common goal and form their strategies based on cooperative interaction, and from the other side, have different competencies and individual characteristics that determine their specialization and efficiency. Thus, the problem of a heterogeneous coalitions structure forming in a network-centric group control system has a game and multicriteria character, and in this aspect has not been sufficiently investigated. *Purpose of research* the structural-goal adaptation intelligent technology development of network-centric group control systems based on multi-criteria game-theoretic heterogeneous coalitions formation models. *Results* A methodology has been developed for a heterogeneous coalition of mobile robots formation based on a hybrid model of multicriteria efficiency estimation and clustering of group control network-centric system infrastructure elements. A hybrid model includes graph and game components. The graph component characterizes the communication properties of infrastructure elements, and the game component characterizes the possibility of cooperation and the availability for each coalition of relevant necessary a competencies set for the task solving. The multicriteria clustering technique is based on the efficiency index of network-centric system infrastructure elements calculating. A library of algorithms for vector effectiveness indicator components of the of a

V. A. Serov (✉) · Y. Yu. Kurilenko
MIREA—Russian Technological University, Moscow, Russia
e-mail: serov_v@mirea.ru

Y. Yu. Kurilenko
e-mail: jlggi@list.ru

E. M. Voronov
Bauman Moscow State Technical University, Moscow, Russia
e-mail: emvoronov@mail.ru

I. L. Tarasova and B. A. Kulik (eds.), *Smart Electromechanical Systems*, Studies in Systems, Decision and Control 544, https://doi.org/10.1007/978-3-031-64277-7_7

network-centric system elements calculating based on network centrality and game-theoretic centrality metrics has been developed. A heterogeneous coalition formation can be considered as one of the tools for mobile robots social behavior models creating in group control tasks *Practical significance* The applied problem of an ad hoc heterogeneous coalition forming in a network-centric system has been solved. To calculate the vector indicator effectiveness components of the infrastructure elements of a network-centric system, the network metric of betweenness centrality and the game-theoretic metric of centrality by the Shapley vector were used. The problem of a network-centric system information infrastructure critical elements identifying from the point of view of ensuring its security has been solved.

Keywords Network-centric system · Group control · Structural-goal adaptation · Heterogeneous coalition · Multicriteria clustering · Shapley vector · Betweenness centrality · Social behavior model

1 Introduction

The development and implementation of smart electromechanical systems (SEMS) group control systems (GCS) is one of the priority research areas of worldwide. In [1–13] the voluminous thematics and researches results covering the main actual the group control problem aspects are presented, which show that one of the group control methodology key tasks is the models and methods development of GCS situational analysis and structural adaptation, providing the possibility of solving a wide class of tasks under the external environment uncertainty.

The numerous bibliographies [5, 12–19] analysis allows us to conclude that game approaches are promising, allowing for various types of uncertain factors: purpose uncertainty, the external environment uncertainty and conflict uncertainty in the models and methods development of structural adaptation of GCS under of uncertainty, which predetermines their universal character. At the same time, within the game approaches framework, structural adaptation mechanisms can naturally be implemented in the form of self-organization models based on the heterogeneous coalitions formation of SEMS-players who are members of the GCS and are focused on complex tasks solving corresponding to the current situation. The heterogeneous coalitions formation allows to unite the capabilities of individual specialized SEMS-players and to obtain a synergistic effect.

Currently, the network-centric group control (NCGC) and information processing methodology is being actively introduced into the theory and practice of solving group control problems [15, 20–29]. NCGS-systems are the next stage in the evolution of GCS and are increasingly being used in applications such as mobile robotics, unmanned aerial vehicles, remote monitoring systems, military, etc.

The network-centric control paradigm is based on the following main principles:

- formation of a unified network information and communication space;
- collection, integration, and up-to-date maintenance of heterogeneous information by using the maximum number of available sources of primary and operational information;
- formation of an environment of situational analysis, forecasting, decision-making and control in real time.

The above principles implementation within the game-theoretic approach framework makes it possible to build an NCGC-systems structural adaptation intelligent model [13–15], which provides the opportunity:

- to form and change the heterogeneous coalitions composition and coalition structure on a set of SEMS-players based on situational analysis and a subject area ontological model;
- taking into account the various types of conflict interaction between coalitions within the coalition structure;
- combining of conflict equilibrium principles in order to form stable and effective SEMS group control strategies;
- optimal communication resources distribution between coalitions in real time.

The task of a heterogeneous coalition forming of SEMS-players is one of the key ones, because its solution allows you to implement a self-organization mechanism in the NCGC-system. At the same time, the evaluation of each coalition effectiveness should take into account functional and communication aspects. The functional aspect characterizes the specialization of the coalition and its importance, determined by the set of its members competencies. The communication aspect characterizes the information connectivity of the coalition participants, as well as the integration degree of the coalition into the information infrastructure of the NCGC-system.

To estimate the functional capabilities of the heterogeneous coalition participants and the entire coalition, in [30–38] a game-theoretic concept of centrality is being developed based on the representation of the SEMS group interaction model as a cooperative game in the form of a characteristic function. The Shapley value, Basnzaf value, Owen value, Myerson value, as well as their modifications are considered as the main game-theoretic metrics of centrality.

The communication capabilities of the network-centric infrastructure participants (elements) and coalitions are evaluated using network centrality metrics. The network centrality metrics concept uses a graph model of a network-centric system and is considered in [39–47], where the main metrics are being investigated: degree centrality, eigenvector centrality, centrality by Katz (generalized eigenvector centrality), betweenness centrality, closeness centrality, decay centrality, PageRank centrality, percolation centrality, Each of the network centrality metrics has its own interpretation and characterizes the of control degree of an NCGC-system element over information flows.

The analysis of the centrality metrics shows that for an objective assessment of heterogeneous SEMS coalitions effectiveness, it is necessary to use a vector criterion that takes into account the functional and communication characteristics of the SEMS included in the of the NCGC-system infrastructure. Thus, the task of a heterogeneous SEMS coalition forming has a multi-criteria nature, and is unsufficiently investigated in this context.

2 The Problem Statement of a SEMSs Coalition Forming in a Network-Centric System

A hybrid model of network-centric infrastructure is considered in the form

$$\Gamma = \langle \mathbf{N}, \mathbf{A}, \nu(\mathbf{K}) \rangle. \tag{1}$$

The hybridity property (1) means its dual character. From the one side, the infrastructure Γ is an undirected weighted graph, where $\mathbf{N} = \{\overline{1,n}\}$ is the nodes set of the graph; $\mathbf{A}$ is the adjacency matrix of the graph, characterizing the relationships between the nodes of the graph and their intensity. On the other side, within the infrastructure (1) framework, it is assumed that the cooperation between the nodes (participants) is possible to achieve common goals by forming coalitions $\mathbf{K} \subset \mathbf{N}$. That is, (1) is also a cooperative game with a characteristic function $\nu(\mathbf{K})$. The graph component of the model (1) characterizes the communication capabilities of infrastructure objects, and the gaming component is the presence of each coalition association of participants (players) $\mathbf{K}$ with a set of relevant competencies necessary for set tasks completing.

We assume that the each participant significance, in the sense of its influence on the entire infrastructure Γ, is estimated by a vector criteria $\mathbf{F}(i) = [f_1(i), f_2(i)]^T$, where $f_1(i)$ characterizes the communication properties of the participant $i \in \mathbf{N}$, and $f_2(i)$—the significance of the participant, taking into account his specialization and the possibility of forming coalitions with other participants in the set task completing.

The component $f_1(i)$ has a sense of normalized betweenness centrality [47] and is calculated by the formula:

$$f_1(i) = \frac{2}{(|\mathbf{N}|-1)(|\mathbf{N}|-2)} \left(\sum_{(k,j)\in \mathbf{S}(i)} \frac{|\mathrm{M}_{kij}|}{|\mathrm{M}_{kj}|} \right). \tag{2}$$

In (2), the following notations are accepted: $\mathbf{S}(i)$—the set of all possible of graph nodes paired combinations (1) on the set $(\mathbf{N}\backslash i)$; M_{kij}—a shortest routes set for a paired combination of nodes (k,j) passing through the node i; M_{kj}—a shortest routes set for the nodes (k,j) paired combination; $\mathrm{M}_{kij} \subseteq \mathrm{M}_{kj}$.

The component $f_2(i)$ has the meaning of the Shapley value [31, 46] and is calculated using the formula:

$$Sh(i) = \sum_{\substack{\mathbf{K} \subseteq \mathbf{N}, \\ i \in \mathbf{K}}} \frac{(|\mathbf{K}| - 1)!(|\mathbf{N}| - |\mathbf{K}|)!}{|\mathbf{N}|!} (v(\mathbf{K}) - v(\mathbf{K} \backslash i)), \tag{3}$$

where $v(\mathbf{K})$ is the characteristic function of the cooperative game in the hybrid model (1).

Let's form a function called the participant i performance index:

$$\Phi(i) = \frac{1}{\left(1 + \frac{b_i}{|\mathbf{N}|-1}\right)^{\psi}}, \tag{4}$$

where ψ is some positive integer; b_i—the number of participants $j \in \mathbf{N} \backslash i$ for whom the condition

$$\mathbf{F}(j) \geq \mathbf{F}(i) \tag{5}$$

is met.

The effectiveness of the coalition $\mathbf{K}$ will be evaluated by a vector criterion $\mathbf{H}(\mathbf{K}) = [h_1(\mathbf{K}), h_2(\mathbf{K})]^T$, the components of which are given in the form:

$$h_1(\mathbf{K}) = v(\mathbf{K}), \tag{6}$$

$$h_2(\mathbf{K}) = \sum_{i \in \mathbf{K}} \Phi(i). \tag{7}$$

Thus, a $r_{\mathbf{K}}$-dimensional optimal coalition $\mathbf{K}$ forming problem statement in a network-centric system is formalized as a discrete multi-criteria optimization problem

$$\mathbf{H}(\mathbf{K}) \to \max_{\mathbf{K} \subset \mathbf{N}^{\mathbf{K}}}. \tag{8}$$

In task (8), $h_1(\mathbf{K})$ is defined in the form (6), interpreted as a participants coalition $\mathbf{K}$ readiness degree to perform the task and characterizes the coalition competence as a single player; $h_2(\mathbf{K})$ is defined in the form (7) and characterizes the $r_{\mathbf{K}}$-dimensional coalition $\mathbf{K}$ importance on a coalitions set $\mathbf{N}^{\mathbf{K}}$:

$$\mathbf{N}^{\mathbf{K}} = \{\mathbf{K} \subset \mathbf{N} | \, |\mathbf{K}| = r_{\mathbf{K}} \}. \tag{9}$$

3 The Algorithm for the Betweenness Centrality Index Calculating for Nodes of a Weighted Graph

To calculate the betweenness centrality, it is required for each node of a weighted undirected graph to determine the number of shortest paths passing through it, and constructed for all graph nodes paired combinations. To do this, it is necessary for each pair $(i,j) \in \mathbf{S}$ to form a shortest routes set M_{ij} on the set $\mathbf{S}$ of all nodes paired combinations. The number of all possible nodes paired combinations on the graph

$$|\mathbf{S}| = \mathrm{C}_n^2 = \frac{n(n-1)}{2}.$$

Therefore generally, it is necessary to solve $|\mathbf{S}|$ problems of a shortest route finding on the graph. For calculations acceleration, it is proposed to form a set of tasks $\mathbf{P} = \left\{\mathrm{P}_i, i = \overline{1,(n-1)}\right\}$ on the set $\mathbf{S}$ of all possible paired combinations. The task P_i is formulated as follows. It is required to find all the shortest paths from a fixed node i to the graph nodes with numbers $j = \overline{(i+1),n}$. That is, within the problem P_i framework, it is necessary to solve $(n - i)$ tasks of the shortest route finding for the corresponding nodes paired combinations. Then the tasks quantity of the shortest route finding for various nodes paired combinations, which must be solved on the whole set $\mathbf{P}$ is calculated as:

$$L = \sum_{i=1}^{n} (n-i) = \frac{n(n-1)}{2} = |\mathbf{S}|.$$

3.1 The Genetic Algorithm for the Task P_i Solving

When developing a genetic algorithm for solving the task P_i, a library of genetic algorithms for multicriteria optimization (GAMO) was used [48]. Each chromosome $\mathbf{h}(i)$ contains possible routes $(i \to j)$ from the current node i to the nodes of the graph with numbers $j = \overline{(i+1), n}$. Integer coding is used. The structure $\mathbf{h}(i)$ is shown in Fig. 1 (in green) and can be represented as

$$\mathbf{h}(i) = \left\{\mathbf{h}_j(i) \middle| j = \overline{(i+1),n}\right\}, \tag{10}$$

where $\mathbf{h}_j(i) = \pi_j(\mathbf{N}\backslash i)$ is the elements permutation of the set $(\mathbf{N}\backslash i)$ corresponding to the section $\mathbf{h}_j(i)$; $\mathbf{\Pi}(\mathbf{N}\backslash i)$—a set of possible permutations $\pi(\mathbf{N}\backslash i)$; $|\mathbf{\Pi}(\mathbf{N}\backslash i)| = (n-1)!$. The section $\mathbf{h}_j(i)$ contains a possible route $s_j = (i \to j)$ (in yellow). Chromosome $\mathbf{h}(i)$ parameters: $l_{\mathbf{h}(i)} = (n-1)(n-i)$—the length of the chromosome; $l_{\mathbf{h}_j(i)} = (n-1)$—the length of the section $\mathbf{h}_j(i)$.

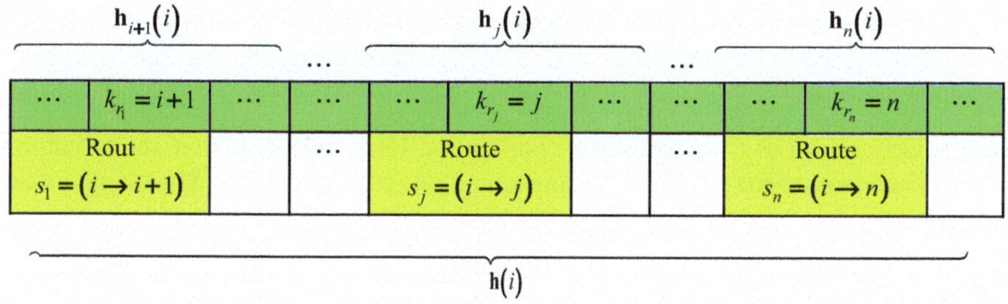

Fig. 1 The chromosome $\mathbf{h}(i)$ structure for the task P_i solving

The chromosome $\mathbf{h}(i)$ is evaluated by a vector function

$$\mathbf{F}(\mathbf{h}(i)) = \left[f_j\left(\mathbf{h}_j(i)\right), j = \overline{(i+1),n}\right]^T \in \mathbf{E}^{n-i}. \tag{11}$$

In (11), the component $f_j\left(\mathbf{h}_j(i)\right)$ value characterize the length of the route $s_j = (i \to j)$ and is calculated as

$$f_j\left(\mathbf{h}_j(i)\right) = \sum_{s=1}^{(r_1|k_{r_1}=j)} d(i,k_s), \tag{12}$$

where $d(i,k_s)$ is the length of the edge $(i \to k_s)$.

Thus, the task P_i of finding all shortest paths from a fixed node i to graph nodes with numbers $J = \overline{(i+1), n}$, can be formalized as a multicriteria optimization problem:

$$\mathbf{F}(\mathbf{h}(i)) \to \min_{\mathbf{h}(i)\in\mathbf{H}(i)}, \tag{13}$$

where the set of chromosome $\mathbf{h}(i)$ acceptable values is given as a Cartesian product:

$$\mathbf{H}(i) = \mathbf{\Pi}^{(n-i)}(\mathbf{N}\backslash i). \tag{14}$$

However, due to the fact that the vector objective function $\mathbf{F}(\mathbf{h}(i))$ has the form (11), the set of Pareto-optimal solutions to the problem (13), (14) consists of a single point. Therefore, problem (13), (14) can be formalized as a problem of separable programming:

$$G(\mathbf{h}(i)) = \sum_{j=i+1}^{n} f_j\left(\mathbf{h}_j(i)\right) \to \min_{\mathbf{h}(i)\in\mathbf{H}(i)}, \tag{15}$$

where the functions $f_j\left(\mathbf{h}_j(i)\right)$ are given in the form (12), and the set $\mathbf{H}(i)$ is given in the form (14).

The genetic algorithm for all shortest routes finding from a given node i to the nodes of the graph with numbers $j = \overline{(i+1), n}$, can be represented as the following sequence of steps.

Step 1. Generation of a chromosomes population $\tilde{\mathbf{H}}(i) \subset \mathbf{H}(i)$, where the structure of the chromosome $\mathbf{h}(i) \in \tilde{\mathbf{H}}(i)$ is represented as (10), and the set $\mathbf{H}(i)$ is given as (14).

Step 2. A suitability estimation of all individuals $\mathbf{h}^k(i) \in \tilde{\mathbf{H}}(i)$ using the fitness function

$$\Phi\left(\mathbf{h}^k(i)\right) = \frac{1}{\left(1 + \frac{b_k(i)}{\left|\tilde{\mathbf{H}}(i)\right|-1}\right)^{\psi}}, \tag{16}$$

where $b_k(i)$ is the number of individuals $\tilde{\mathbf{h}}$ in the population $\tilde{\mathbf{H}}$ for which the condition

$$G\left(\tilde{\mathbf{h}}(i)\right) < G\left(\mathbf{h}^k(i)\right), \tag{17}$$

is met; ψ—parameter that affects the convergence rate of the genetic algorithm as a whole through the selection operator. As the ψ increases, the genetic algorithm rate of convergence increases. However, the influence of the deterministic component of the algorithm increases while simultaneously the influence of the stochastic component is weaking. This can lead to premature termination of the algorithm and obtaining local optimal solutions.

Step 3. A parents population formation. The selection operator, based on the combined implementation of tournament selection and roulette methods is used.

Step 4. A descendants population formation. In solving this problem, the PMX crossover operator (Partially Mapped Crossover) has shown high efficiency.

Step 5. Elite solutions population formation. Individuals for whom the fitness function takes the value 1 are copied into an array of elite solutions.

Step 6. A stop condition checking. If the stop condition is fulfilled, then go to step 7. Otherwise, go to step 2.

Step 7. Based on the elite solutions population, we form sets of optimal routes M_{ij}, $j = \overline{(i+1), n}$.

3.2 A Calculation of Node Centrality by Betweenness

Step 1. Solve the set of tasks $\mathbf{P} = \left\{P_i, i = \overline{1, (n-1)}\right\}$. All optimal routes obtained as a set $\mathbf{P} = \left\{P_i, i = \overline{1, (n-1)}\right\}$ of optimization problems solving result can be represented as a data structure $\mathrm{M} = \{\mathrm{M}_{kj}, (k,j) \in \mathbf{S}\}$, where $\mathrm{M}_{kj} = \{\mu_{kj}\}$ is the set of shortest routes for each paired combination of nodes $(k,j) \in \mathbf{S}$. The generated data structure M is shown in the form of Table 1.

Step 2. For a fixed i, we construct the set $\mathbf{S}(i)$ of all possible paired combinations on the set of nodes $(\mathbf{N}\backslash i)$; $|\mathbf{S}(i)| = C^2_{(n-1)} = \frac{(n-1)(n-2)}{2}$. The data structure $\mathrm{M}(i) = \{\mathrm{M}_{kj}, (k,j) \in \mathbf{S}(i)\}$ corresponds to the set $\mathbf{S}(i)$. As follows from Table 1, $\mathrm{M}(i)$ is obtained from M by crossing out the row and column with the number i (marked in grey).

Step 3. We calculate a normalized centrality by betweenness of node i based on the data structure $\mathrm{M}(i)$:

$$B(i) = \frac{1}{C^2_{(n-1)}} \left(\sum_{(k,j) \in \mathbf{S}(i)} \frac{|\mathrm{M}_{kij}|}{|\mathrm{M}_{kj}|} \right). \tag{18}$$

In (18) M_{kij}—the set of shortest routes for a paired combination of nodes $(k,j) \in \mathbf{S}(i)$ passing through the node i; $\mathrm{M}_{kij} \subseteq \mathrm{M}_{kj}$.

Remark. The problem decomposition of all shortest routes finding in the form of a set of tasks **P** with subsequent formation of a data structure M is the basis for a parallel algorithm constructing for calculating the centrality index by betweenness.

Table 1 Data structure **M**

	1		…	k	…	i	…	j	…	n
1	×		…	$\mathbf{M}_{1k}$	…	$\mathbf{M}_{1i}$	…	$\mathbf{M}_{1j}$	…	$\mathbf{M}_{1n}$
…	×		×	…	…	…	…	…	…	…
k	×		×	×	…	$\mathbf{M}_{ki}$	…	$\mathbf{M}_{kj}$	…	$\mathbf{M}_{kn}$
…	×		×	×	×	…	…	…	…	…
i	×		×	×	×	×	…	$\mathbf{M}_{ij}$	…	$\mathbf{M}_{in}$
…	×		×	×	×	×	×	…	…	…
j	×		×	×	×	×	×	×	…	$\mathbf{M}_{jn}$
…	×		×	×	×	×	×	×	×	…
n	×		×	×	×	×	×	×	…	×

4 The Algorithm for the Centrality Index Calculating Based on the Shapley Vector

The participant $i \in \mathbf{N}$ centrality index based on the Shapley vector is calculated as (3). In this case the Shapley vector components values significantly are depended of the coalition characteristic function calculating mechanism.

It is proposed to interpret the characteristic function as the readiness degree of a coalition to be used for solving the task. As the initial data for calculating the participant coalition characteristic function we'll use the signature of each SIMS-member of the coalition and the adjacency matrix of the original graph. The signature contain data on competencies and technical characteristics that determine the individual characteristics of the SEM and the whole coalition **K**, and are presented in the form of Table 2.

The elements of Table 2 have the following meaning:

$$C_p^i = \begin{cases} 1, \text{ if SEM}_i \text{ has competention } C_p, \\ 0, \text{ otherwise.} \end{cases}; \quad S_q^i = \begin{cases} 1, \text{ SEMS}_i \text{ has characteristic } S_q, \\ 0, \text{ otherwise.} \end{cases}$$

The signature of the coalition **K** is a $(P + Q)$-bit string, which is formed in accordance with the following rule.

$$C_p^{\mathbf{K}} = \sum_{i \in \mathbf{K}} C_p^i; \quad S_q^{\mathbf{K}} = \sum_{i \in \mathbf{K}} S_q^i \tag{19}$$

The algorithm for characteristic function calculating includes the following steps.

Step 1. Checking the condition Req1 met:

$$\text{Re}q1\text{: }\{\mathbf{K} - \text{ connected subgraph.}$$

If the condition Req1 is met, then go to step 2. Otherwise, we assume $v(\mathbf{K}) = 0$ and go to step 4.

Table 2 Signatures of SEMS as participants of a cooperative game

Participants $i \in \mathbf{N}$	Competentions					Characteristics				
	C_1^i	...	C_p^i	...	C_P^i	S_1^i	...	S_q^i	...	S_Q^i
1	0	0	...	1	1	0	...	1	...	0
...	...	...	...	...	...	...	...	...	...	...
i	1	...	1	...	0	1	...	0	...	1
...	...	...	...	...	...	...	...	...	...	...
n	0	...	1	...	0	1	...	0	...	1
K	$C_1^{\mathbf{K}}$	...	$C_p^{\mathbf{K}}$	...	$C_P^{\mathbf{K}}$	$S_1^{\mathbf{K}}$	...	$S_q^{\mathbf{K}}$	...	$S_Q^{\mathbf{K}}$

Step 2. Checking the condition Req2 met:

$$\text{Re}q2\text{:}\begin{cases} C_p^{\mathbf{K}} \neq 0, & \forall p = \overline{1,P}; \\ S_q^{\mathbf{K}} \neq 0, & \forall q = \overline{1,Q}. \end{cases}$$

If the condition Req2 is met, then go to step 3. Otherwise, we assume $v(\mathbf{K}) = 0$ and go to step 4.

Step 3. The characteristic function calculation in the form:

$$v(\mathbf{K}) = \sum_{p=1}^{P} C_p^{\mathbf{K}}. \tag{20}$$

Step 4. The algorithm stopping.

5 The Genetic Algorithm for an Optimal Coalition Forming

To solve the problem of an optimal coalition forming (6)–(9), the GAMO-library was used [48]. Each chromosome **h** contains a composition of $r_{\mathbf{K}}$-dimensional coalition **K** and is represented as:

$$\mathbf{h} = \left(\mathbf{g}_1, \ldots, \mathbf{g}_{r_{\mathbf{K}}}\right), \tag{21}$$

where $\mathbf{g}_i$ is the binary Gray code (BGC) of the coalition participant $i \in \mathbf{K}$ (the number of the corresponding node of the graph). Chromosome **h** parameters: $l_{\mathbf{h}} = r r_{\mathbf{K}}$—the length of the chromosome, where r is the bit depth of the BGC of $\mathbf{g}_i$. The parameter r value is selected from the inequality meting condition: $|\mathbf{N}| \leq 2^r - 1$.

Chromosome (21) is evaluated by a vector function $\mathbf{H}(\mathbf{K}) = [h_1(\mathbf{K}), h_2(\mathbf{K})]^T$ with components in the form (6), (7), the calculation algorithms of which have been discussed above. At the same time, in the algorithm for calculating the Shapley vector, when checking the subgraph connectivity condition Req1, an additional condition is used that simplifies the calculation of the characteristic function: if after decoding $\mathbf{g}_i$ an inequality $i > |\mathbf{N}|$ is satisfied, then the connectivity condition of the subgraph **K** is not met, and we assume $v(\mathbf{K}) = 0$.

6 Computational Experiment and Results Analysis

The hybrid infrastructure model of a network-centric group control system is presented in the form (1) including graph and game components.

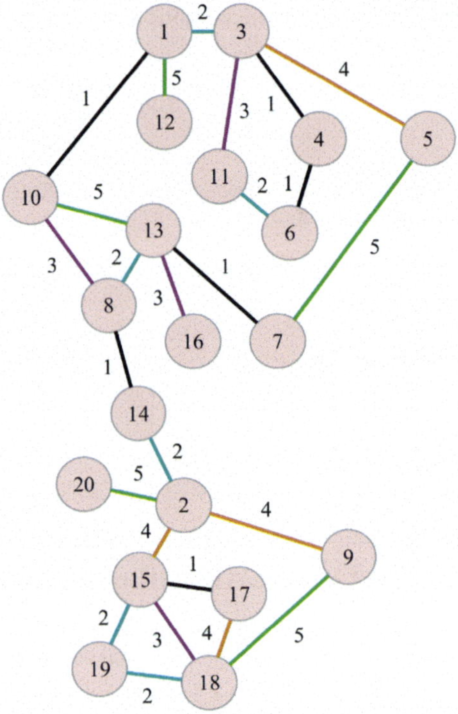

Fig. 2 The network-centric system infrastructure graph model

The network-centric system model graph component is shown in Fig. 2 as a weighted undirected graph, where $|\mathbf{N}| = 20$.

The adjacency matrix of the graph is presented in the form of Table 3.

The network-centric system model game component is defined as a cooperative game in the form of a characteristic function. The task model to be performed is characterized by the signature shown in Table 4 as a set of necessary competencies and characteristics.

The signatures of SEMs-participants are shown in Table 5.

It is required to form a heterogeneous coalition $\mathbf{K}$, $r_{\mathbf{K}} = 4$, to completing the formulated task.

The results of the Shapley vector calculating, centrality by betweenness, and multi-criteria ranking of SEMSs-elements of a network-centric system based on calculating the efficiency index are presented in Table 6, and are further used as initial information for vector evaluation of the coalitions effectiveness in solving the forming an optimal coalition problem.

The problem of an optimal coalition forming is solved in the form (6)–(8) using the GAMO-library. The following basic parameters and characteristics of the genetic algorithm are used:

- BGC $\mathbf{g}_i$, $i = \overline{1,4}$, bit depth: $r = 5$;
- Chromosome $\mathbf{h}$ length: $l_{\mathbf{h}} = r r_{\mathbf{K}} = 20$;

Table 3 The adjacency matrix of the network-centric system infrastructure graph model

Participant $i \in \mathbf{N}$	1	2	3	4	5	6	7	8	9	10	11	12	13	14	15	16	17	18	19	20
1	0	0	2	0	0	0	0	0	0	1	0	5	0	0	0	0	0	0	0	0
2	0	0	0	0	0	0	0	0	4	0	0	0	0	2	4	0	0	0	0	5
3	2	0	0	1	4	0	0	0	0	0	3	0	0	0	0	0	0	0	0	0
4	0	0	1	0	0	1	0	0	0	0	0	0	0	0	0	0	0	0	0	0
5	0	0	4	0	0	0	5	0	0	0	0	0	0	0	0	0	0	0	0	0
6	0	0	0	1	0	0	0	0	0	0	2	0	0	0	0	0	0	0	0	0
7	0	0	0	0	5	0	0	0	0	0	0	0	1	0	0	0	0	0	0	0
8	0	0	0	0	0	0	0	0	0	3	0	0	2	1	0	0	0	0	0	0
9	0	4	0	0	0	0	0	0	0	0	0	0	0	0	0	0	0	5	0	0
10	1	0	0	0	0	0	0	3	0	0	0	0	5	0	0	0	0	0	0	0
11	0	0	3	0	0	2	0	0	0	0	0	0	0	0	0	0	0	0	0	0
12	5	0	0	0	0	0	0	0	0	0	0	0	0	0	0	0	0	0	0	0
13	0	0	0	0	0	0	1	2	0	5	0	0	0	0	0	3	0	0	0	0
14	0	2	0	0	0	0	0	1	0	0	0	0	0	0	0	0	0	0	0	0
15	0	4	0	0	0	0	0	0	0	0	0	0	0	0	0	0	1	3	2	0
16	0	0	0	0	0	0	0	0	0	0	0	0	3	0	0	0	0	0	0	0
17	0	0	0	0	0	0	0	0	0	0	0	0	0	0	1	0	0	4	0	0
18	0	0	0	0	0	0	0	0	5	0	0	0	0	0	3	0	4	0	2	0
19	0	0	0	0	0	0	0	0	0	0	0	0	0	0	2	0	0	2	0	0
20	0	5	0	0	0	0	0	0	0	0	0	0	0	0	0	0	0	0	0	0

Table 4 Signature of the task model be performed

Competencies					Characteristics	
C_2^i	C_2^i	C_3^i	C_4^i	C_5^i	S_1^i	S_2^i
1	1	1	1	1	1	1

- the size of the chromosome population: $|\tilde{\mathbf{H}}| = 200$;
- the fitness of all individuals $\mathbf{h}^k \in \tilde{\mathbf{H}}$ is evaluated using the fitness function in form (16), where b_k is the number of individuals $\tilde{\mathbf{h}}$ in the population $\tilde{\mathbf{H}}$, for which the vector inequality $\mathbf{H}(\tilde{\mathbf{h}}) \geq \mathbf{H}(\mathbf{h}^k)$ met, a parameter $\psi = 2$;
- type of selection operator: a tournament scheme and roulette combining;
- type of crossover operator: single point within the boundaries of each $\mathbf{g}_i$, $i = \overline{1, 4}$;
- crossover probability: $p_c = 0.90$;
- type of mutation: single-point, within the boundaries of one randomly selected $\mathbf{g}_i$;
- the probability of mutation: $p_b = 0.05$.

Table 5 The signatures of SEMs-participants

Participants $i \in \mathbf{N}$	Competentions					Characteristics	
	C_1^i	C_2^i	C_3^i	C_4^i	C_5^i	S_1^i	S_2^i
1	1	0	0	0	0	0	0
2	1	0	1	0	0	0	0
3	1	0	0	1	0	0	0
4	1	0	0	0	1	0	0
5	1	0	0	0	0	1	0
6	0	1	0	0	1	0	0
7	0	1	0	1	0	0	0
8	0	1	0	1	0	0	0
9	0	0	1	0	0	0	0
10	0	0	1	0	1	0	0
11	0	0	1	0	0	1	1
12	0	0	0	1	0	0	0
13	0	1	0	1	0	0	0
14	1	0	0	0	0	0	0
15	0	0	0	0	1	0	0
16	0	1	0	0	0	1	0
17	0	0	1	0	0	0	1
18	0	0	0	1	1	1	0
19	0	0	1	0	0	1	0
20	0	1	0	0	0	0	1

The Pareto optimal solutions set of the problem (6)–(8) $\mathcal{K} = \left\{\mathbf{K}_1^{opt}, \mathbf{K}_2^{opt}, \mathbf{K}_3^{opt}\right\}$ is presented in Table 7.

The solution $\mathbf{K}_3^{opt}$ is formally Pareto-optimal. However, it does not met the requirement of readiness for use, because $v\left(\mathbf{K}_3^{opt}\right) = 0$. Therefore, coalitions should be considered as optimal solutions $\mathbf{K}_1^{opt}$, $\mathbf{K}_2^{opt}$.

Other approaches to solving the optimal heterogeneous coalition forming problem based on a hybrid model (1) are possible. Communication and game components can be simultaneously taken into account within the framework of the following problem statements.

Problem statement 1. The optimization problem is solved according to the criterion of the coalition's readiness for use:

$$v(\mathbf{K}) \rightarrow \max_{\mathbf{K} \subset \mathbf{N}^{\mathbf{K}}} . \tag{22}$$

In this formulation, the form of the characteristic function plays a decisive role. The optimal solution to problem (22) is a coalition $\mathbf{K}^{opt1} = \{3, 4, 6, 11\}$, $v\left(\mathbf{K}^{opt1}\right) = 7$, which is the same as $\mathbf{K}_1^{opt}$ in problem (6)–(8).

Problem statement 2. The problem of coalition forming, consisting of the most significant SEMSs-participants is solved on the basis of ranking according to the efficiency indexes shown in Table 5, which is equivalent to setting the task

Table 6 Shapley value, centrality by betweenness, efficiency index of SEMSs-elements of a network-centric system

№	Shapley value	Centrality by betweenness	Efficiency index ($\psi = 2$)
2	6.222	0.491	1.0
8	0.735	0.515	1.0
1	5.659	0.339	0.903
13	6.137	0.222	0.903
3	2.193	0.310	0.819
14	0.267	0.491	0.819
10	0.733	0.339	0.746
7	0.992	0.117	0.682
18	1.555	0.111	0.682
11	1.287	0.099	0.627
9	0.294	0.175	0.578
15	1.235	0.088	0.578
4	1.132	0.0	0.534
6	1.038	0.0	0.496
5	0.308	0.082	0.429
17	0.519	0.0	0.376
19	0.246	0.0	0.295
20	– 0.257	0.0	0.279
16	– 0.525	0.0	0.264
12	– 0.771	0.0	0.250

Table 7 Pareto optimal coalitions

№	$\mathbf{K}_i^{opt}$	$h_1(\mathbf{K}^{opt})$	$h_2(\mathbf{K}^{opt})$
1	{3,4,6,11}	7	2.476
2	{2,15,18,20}	6	2.539
3	{1, 2,8, 13}	0	3.806

$$h(\mathbf{K}) = \sum_{i \in \mathbf{K}} \Phi(i) \to \max_{\mathbf{K} \subset \mathbf{N}^{\mathbf{K}}} . \tag{23}$$

The problem (23) optimal solution is the coalition $\mathbf{K}^{opt2} = \{1, 2, 8, 13\}$, $v(\mathbf{K}^{opt2}) = 0$, which is the same, as optimal coalition $\mathbf{K}_3^{opt}$ in problem (6)–(8).

Problem statement 3. The coalition formation of the most significant SEMS-participants in the infrastructure (1) in accordance n_j Shapley vector is equivalent to problem:

$$h(\mathbf{K}) = \sum_{i \in \mathbf{K}} Sh(i) \to \max_{\mathbf{K} \in \mathbf{N}^{\mathbf{K}}} . \tag{24}$$

The optimal solution of problem (24) is a coalition $\mathbf{K}^{opt3} = \{1, 2, 3, 13\}$, $h(\mathbf{K}^{opt3}) = 20.211$. However, the coalition $\mathbf{K}^{opt3}$ is unacceptable because $v(\mathbf{K}^{opt3}) = 0$. This is due to the fact that the components of the Shapley vector characterize the average significance of each SEMS-participant in infrastructure (1) for all possible coalitions $\mathbf{K}$ of dimension $1 \le |\mathbf{K}| \le |\mathbf{N}|$. At the same time, within the framework of a separate coalition, the importance of each participant may be minimal, which leads to a minimum value of the function $v(\mathbf{K}) = 0$ within the framework of the set task.

Summarizing the above, we can conclude that the a coalition forming problem statement (6)–(8) makes it possible to build a set of effective solutions, that provide more meaningful information from the point of view supporting optimal decision-making compared to other considered problem statements.

7 Conclusion

The article develops a methodology for the SEMS heterogeneous coalition formation based on a hybrid infrastructure model of a network-centric group control system. The hybrid model includes graph and game components, where the graph component characterizes the communication capabilities of infrastructure elements, and the game component characterizes the possibilities of cooperation between participants by forming coalitions to complete the set task.

The elements effectiveness indicators set of the network-centric group control system and their heterogeneous coalitions was formed, for the calculation of which the network metric of centrality by betweenness and the game-theoretic metric of centrality by the Shapley vector were used. The interpretation of the characteristic function is proposed as the degree of readiness of the coalition to be used to solve the task. The signatures of the SEMS members of the coalition and the adjacency matrix of the original graph, which determines the connectivity characteristic of the coalition, are used as initial data for calculating the characteristic function of the coalition. The signatures contain data on competencies and technical characteristics that determine the individual characteristics of SEMS and the whole coalition, which characterized heterogeneity of the coalition being formed.

The problem statement of optimal coalition forming in a network-centric system is formalized as a discrete multicriteria optimization problem. To solve this problem, combined evolutionary computational procedures for calculating the centrality index by betweenness for the nodes of a weighted graph and the optimal heterogeneous coalition formation in a cooperative game in the form of a characteristic function have been developed, expanding the capabilities of the GAMO-library [48]. The proposed approach to the heterogeneous coalition formation can be considered as one of the tools for SEMS social behavior models building in group control tasks.

The problem of a heterogeneous coalition of SEMS forming in a network-centric system to perform a specialized task, defined as a signature representing a set of required competencies and technical characteristics of SEMS has been solved. It is shown that the developed methodology, in comparison with other approaches, makes it possible to obtain a more complete set of optimal coalitions with the necessary performance potential.

References

1. Gorodetskiy, A., Tarasova, I. (eds.): Smart Electromechanical Systems. Group Interaction. Studies in Systems, Decision and Control, vol. 174. Springer International Publishing (2019). https://doi.org/10.1007/978-3-319-99759-9
2. Gorodetskiy, A., Tarasova, I. (Eds.): Smart Electromechanical Systems. Situational Control. Studies in Systems, Decision and Control, vol. 261, 290 p. Springer International Publishing (2020). https://doi.org/10.1007/978-3-030-32710-1_9
3. Gorodetskiy, A., Tarasova, I. (eds.): Smart Electromechanical Systems. Behavioral Decision Making. Studies in Systems, Decision and Control, vol. 352, 240 p. Springer International Publishing (2021). https://doi.org/10.1007/978-3-030-68172-2_13
4. Gorodetskiy, A., Tarasova, I. (eds.): Smart Electromechanical Systems. Recognition, Identification, Modeling, Measurement Systems, Sensors. Studies in Systems, Decision and Control, vol. 419, 275 p. Springer International Publishing (2022). https://doi.org/10.1007/978-3-030-97004-8_10
5. Obnosov, B.V., Voronov, E.M., Mikrin, E.A., Serov, V.A., et al.: Stabilization, Guidance, Group Control, and System Modeling of Unmanned Aerial Vehicles. Modern Approaches and Methods, vol. 2, p. 520. Publishing House of Bauman Moscow state technical University, Moscow (2018). ISBN 978-5-7038-5058-9
6. Pshikhopov, V.Kh.: Group Control of Moving Objects in Uncertain Environments, p. 305. FIZMATLIT (2015). ISBN 978-5-9221-1674-9
7. Kalyaev, I.A.: Models and Algorithms of Collective Control in Groups of Robots, p. 280 (Kalyaev, I.A., Gayduk, A.R., Kapustyan, S.G., eds.). FIZMATLIT, Moscow (2009). ISBN 978-5-9221-1141-6
8. Karpov, I.E., Karpova, I.P., Kulinich, A.A.: Social Communities of Robots, p. 352. LENAND, Moscow (2019). ISBN 978-5-9710-5757-4
9. Navarro, I., Matha, F.: Survey of collective movement of mobile robots. Int. J. Adv. Robot. Syst. **10**(73) (2013)
10. Bertuccelli, L., Choi, H-L., Cho, P., How, J.: Real-time multi-UAV task assignment in dynamic and uncertain environments. In: American Institute of Aeronautics and Astronautics Guidance, Navigation, and Control Conference, Chicago, Illinois, USA, 10–13 Aug 2009. AIAA-2009-5776, pp. 1–16 (2009)
11. Xu, D., Zhang, X., Zhu, Z., Chen, C., Yang, P.: Behavior-based formation control of swarm robots. Math. Probl. Eng. **1**, 1–13 (2014)
12. Ungureanu, V.: Pareto-Nash-Stackelberg Game and Control Theory: Intelligent Paradigms and Applications. Smart Innovation, Systems and Technologies, 579 p. Springer (2017). ISBN 10: 3319751506; ISBN 13: 978-3319751504
13. Serov, V.A., Babintsev, Yu.N., Kondakov, N.S.: Neurocontrol of Multicriteria Conflict Systems, p. 136. MosGU, Moscow (2011). ISBN 978-5-98079-747-8
14. Voronov, E.M., Serov, V.A.: A coordinated stable-effective compromises based methodology of design and control in multi-object systems. In: Gorodetskiy, A., Tarasova, I. (eds.) Studies in Systems, Decision and Control. Smart Electromechanical Systems, vol. 174, pp. 147–157.

Springer International Publishing (2019). https://link.springer.com/book/10.1007/978-3-319-99759-9. https://doi.org/10.1007/978-3-319-99759-9.
15. Serov, V.A., Voronov, E.M., Kozlov, D.A.: Hierarchical neuro-game model of the FANET based remote monitoring system resources balancing. In: Gorodetskiy, A., Tarasova, I. (eds.) Studies in Systems, Decision and Control. Smart Electromechanical Systems. Situational Control, vol. 261, pp. 117–130. Springer International Publishing (2020). https://doi.org/10.1007/978-3-030-32710-1_9
16. Voronov, E.M., Serov, V.A., Kozlov, D.A.: Coalition game model of FANET grouping control based on the method of local threats and counter-threats and swarm-leader model. In: Studies in Systems, Decision and Control, vol. 352, pp. 151–162. Springer International Publishing (2021). https://doi.org/10.1007/978-3-030-68172-2_13
17. Smirnov, A.V., Sheremetov, L.B.: Models of forming coalitions of cooperative agents: state and prospects of research. Artif. Intell. Decis. Making **1**, 36–48 (2011)
18. Klusch, M., Gerber, A.: A Dynamic coalition formation scheme for rational agents. IEEE Intell. Syst., 42–47 (2002)
19. Romero Cortés, J., Sheremetov, L.: Model of cooperation in multi agent systems with fuzzy coalitions. In: Dunin-Keplicz, B., Nawarecki, E. (eds.) From Theory to Practice in Multi-agent Systems (2002)
20. DeCristofaro, M., Lansdowne, C., Schlesinger, A.: Heterogeneous wireless mesh network technology evaluation for space proximity and surface applications. In: SpaceOps 2014 Conference (2014). https://doi.org/10.2514/6.2014-1600
21. Goforth, M., Ratliff, J., Barton, R., Wagner, R., Lansdowne, C.: Avionics architectures for exploration: wireless technologies and human spaceflight. In: 2014 IEEE International Conference on Wireless for Space and Extreme Environments (WiSEE) (2014). https://doi.org/10.1109/WiSEE.2014.6973073
22. Diane, S.A.K., Iskhakov, A.Y., Iskhakova, A.O.: Network-centric motion control algorithm for a group of mobile robots. Model. Optim. Inf. Technol. **10**(1) (2022). Available from: https://moitvivt.ru/ru/journal/pdf?id=1146. https://doi.org/10.26102/2310-6018/2022.36.1.026.. (In Russian)
23. Mulyukha, V.A., Guk, M.Yu., Zaborovsky, V.S.: Supervisory network-centric control system for robotic objects. Robototehnika and Technicheskaya Kibernetika **3**(12), 42–47 (2016)
24. Makarenko, S.I., Oleynikov, A.Y., Chernitskaya, T.E.: Models of interoperability assessment for information systems. Syst. Control Commun. Secur. **4**, 215–245 (2019). https://doi.org/10.24411/2410-9916-2019-10408. (In Russian)
25. Steel, J., Drogemuller, R., Toth, B.: Model interoperability in building information modelling. Softw. Syst. Model. **11**(1), 99–109 (2012)
26. New European Interoperability Framework. Promoting Seamless Services and Data Flows for European Public Administrations, 48 c. Publications Office of the European Union, Luxembourg (2017). https://ec.europa.eu/isa2/sites/isa/files/eif_brochure_final.pdf
27. Sahingoz, O.K.: Networking models in flying ad-hoc networks (FANETs): concepts and challenges. J. Intell. Robot. Systems. – 2014. – T. 74. – №. 1–2. – C. 513–527.
28. Wang, J., Jiang, C., Han, Z., Ren, Y., Maunder, R., Hanzo, L.: Taking drones to the next level: Cooperative distributed unmanned-aerial-vehicular networks for small and mini drones. TIEEE Veh. Technol. Mag. **12**(3), 73–82 (2017)
29. Martynova, L.A., Rozengauz, M.B.: Efficient operation of a group of standalone unmanned submersibles in a network-centric system of underwater illumination. Informatsionno-upravliaiushchie sistemy [Information and Control Systems] **3**, 47–57 (2017). https://doi.org/10.15217/issnl684-8853.2017.3.47. (In Russian)
30. Skibski, O., Michalak, T.P., Rahwan, T., Wooldridge, M.: Algorithms for the Shapley and Myerson values in graph-restricted games. In: Adaptive Agents and Multi-agent Systems (2014)
31. Michalak, T., Aadithya, K., Szczepański, P., Ravindran, B., Jennings, N.: Efficient computation of the Shapley value for game-theoretic network centrality. J. Artif. Intell. Res. **46**, 607–650 (2013)

32. Van Campen, T., Hamers, H., Husslage, B., Lindelauf, R.: A new approximation method for the Shapley value applied to the WTC 9/11 terrorist attack. Soc. Netw. Anal. Min. **8**(1) (2018)
33. Lindelauf, R., Hamers, H., Husslage, B.: Game theoretic centrality analysis of terrorist networks: the cases of Jemaah Islamiyah and Al Qaeda. CentER Discussion Paper, №107 (2011)
34. Algaba, E., Prieto, A., Saavedra-Nieves, A.: Rankings in the Zerkani network by a game theoretical approach, 28c (2022). https://arxiv.org/abs/2202.07730v1
35. Algaba, E., Prieto, A., Saavedra-Nieves, A., Hamers, H.: Analyzing the Zerkani network with the Owen value. In: Advances in Collective Decision Making, pp. 225–242. Springer (2023)
36. Myerson, R.B.: Graphs and cooperation in games. Math. Oper. Res. **2**(3), 225–229 (1977)
37. A connectivity game for graphs. Math. Methods Oper. Res. **60**, 453–470 (2004)
38. Narayanam, R., Narahari, Y.: A Shapley value-based approach to discover influential nodes. IEEE Trans. Autom. Sci. Eng. **8**, 130–147 (2011)
39. Fletcher, J.M., Wennekers, T.: From structure to activity: using centrality measures to predict neuronal activity. Int. J. Neur. Syst. (2017). https://doi.org/10.1142/S0129065717500137
40. Newman, M.E.J.: Finding community structure in networks using the eigenvectors of matrices. Phys. Rev. E **74**(3), 036104 (2006). https://doi.org/10.1103/PhysRevE.74.036104
41. Van Mieghem, P.: Graph Spectra for Complex Networks, 346 p. Cambridge University Press, Cambridge, NY (2011)
42. Borgatti, S.P., Everett, M.G.: A graph-theoretic perspective on centrality. Soc. Netw. **28б**, 466–484 (2005)
43. Rochat, Y.: Closeness centrality extended to unconnected graphs: the harmonic centrality index. In: Applications of Social Network Analysis (ASNA 2009) (2009)
44. Boldi, P., Vigna, S.: Axioms for centrality. Internet Math. **10**(3–4) (2014). https://doi.org/10.1080/15427951.2013.865686
45. Piraveenan, M. Percolation centrality: quantifying graph-theoretic impact of nodes during percolation in networks. PLOS ONE **8**(1), 53095 (2013). https://doi.org/10.1371/journal.pone.0053095
46. Szczepański, P., Michalak, T., Rahwan, T.: A new approach to betweenness centrality based on the Shapley value. In: Conference: Proceedings of the 11th International Conference on Autonomous Agents and Multiagent Systems, vol. 1, pp. 239–246 (2012)
47. Brandes, U.: A faster algorithm for betweenness centrality. J. Math. Sociol. **25**, 163–177 (2001)
48. Serov, V.A.: Genetic algorithms of conflict equilibriums-based multicriteria systems control optimization under uncertainty. Vestnik BMSTU. Ser. Instrument Making **4**(69), 70–80 (2007)

Design Ontology for Situational Digital Twins in Modeling Structural Safety Within a Group of Intelligent Robots

Alexander Ya Fridman

Abstract *Problem statement*: The paper analyses some opportunities to use specialized digital twins for situational modeling of a non-stationary, spatially distributed group of SEMS robots by means of the intelligent modeling system previously developed by the author. *Purpose of research*: To increase the flexibility of modeling newly created collectives of robots, the concept of structural safety is presented, which generalizes ways to take into account various known aspects of safety. The relevance of this task is determined by the complication of robots' interactions and the corresponding increase in the possibility of the most dangerous dependent (complex, cascading) failures, as well as the growth in the volume of data available for analysis as a result of the rapid development of the Internet of Things (IoT). *Results:* An ontology has been proposed for designing of situational digital twins aimed at the coordination of robots' interactions, as well as for the preventive safety analytics of such interactions. The novelty of the proposed approach lies in the comprehensive application of expert knowledge at all stages of modeling within a cause-and-effect paradigm in order to synthesize preferred options for implementing organizational structures for groups of robots. *Practical significance*: The presented approach will allow developing a new sector of the IoT, the Internet of Models of Things, where the situational digital twins of different components of systems, and robots in particular, will be used for maintaining coordinated and safe functioning of a team of robots in the course of implementing a joint task.

Keywords SEMS-based robot · Structural safety · Design ontology · Coordination of robots' interactions

A. Y. Fridman (✉)
Institute for Informatics and Mathematical Modelling, Kola Science Centre of the Russian Academy of Sciences, Apatity, Russia
e-mail: fridman@iimm.ru

I. L. Tarasova and B. A. Kulik (eds.), *Smart Electromechanical Systems*, Studies in Systems, Decision and Control 544, https://doi.org/10.1007/978-3-031-64277-7_8

1 Introduction

Simulation of structurally complex objects serves as the only tool for their study, especially at the initial stages of their life cycle, when it is fundamentally impossible to obtain a sufficient amount of information about their characteristics and functioning features, and one has to rely heavily on the knowledge of experts, while at the same time involving mathematical models of components of such objects, if any. In this paper, we continue analyzing control techniques for groups of SEMS robots (GSRs) assigned to implementing a common task [1–4].

This problem is relevant at all stages of GSRs' existence, but the possibilities to model it are growing significantly as design and management methods are improved. A new step of progress in the field of building flexible systems for computerized study of GSRs is associated with the rapid development of the Internet of Things, which significantly increases the amount of information about the characteristics of such objects (usually in the form of time series of values of these characteristics) and makes the issue of correct structuring and application of this information even more pressing.

Tools for modeling complex objects most often reduce the complexity of the model by moving to studying either the probabilistic properties of the object (entropy [5]), mathematical expectations and probabilistic moments of a higher order [6], or the graph properties of the model (connectivity, attainability of states, etc.) [7]. However, there are subject areas in which these approaches to creating a model are unsuitable, since it is necessary to detect specific combinations of values of the characteristics of an object that generate desirable or undesirable trajectories of its behavior. Of course, today's fashionable neural networks (e.g., [8]) with any depth of learning cannot be used here due to the lack of necessary data for such learning. Problems of this type arise in the theory of optimal control, in the analysis of bifurcations of nonlinear systems and in many other areas, but for the analysis and synthesis of GSRs, the safety aspect of their operation is more important, not only within the object itself, but also in its interactions with other objects, especially if they are potentially dangerous and form critical infrastructures [9]. Cause-and-effect models seem to be the only acceptable ones in such cases, but the support systems for these models must allow for their high dimension (number of variables), sufficient to achieve acceptable modelling adequacy. Accordingly, the only way to cope with the complexity of such models is to construct them in the form of hierarchies [10].

The most flexible environment for formalizing knowledge and data about complex systems seems to be a *design ontology*, which, according to [11], should structuredly describe concepts of a subject area, criteria and models applicable for studying an object, methods and scenarios for its operation, ensure effective accumulation and processing of heterogeneous data and knowledge about it. The author argues that design ontology is a formalized description of the knowledge of design subjects regarding the process of designing new or modernizing already known artifacts, including the knowledge about the design object itself and artifacts similar in properties, as well as a thesaurus of the subject domain. Another publication [12] tells that design ontology is considered as a technology for organizing a special type of activity, focused on the development and implementation of design solutions in both the field of material and social world. The enterprise ontology is aimed, first of all, at creating models for organizing activities by tasks, products and territory. A special role is given to corporate classifiers and dictionaries, which are the basis of the ontological repository. Ultimately, the design of enterprise activities comes down to a well-founded technology for creating and implementing the ontological model BEOM (Business Entity Ontological Model).

The subsequent sections of this work propose a solution to the problem of studying the interactions of industrial-natural complexes with using a specialized ontology for designing ant testing the efficiency of new GSRs within the framework of the previously developed theory and system for situational modeling (SSM) [13]. The SSM belongs to the class of simulation modeling systems like Powersim [14] and VisSim [15], but it also incorporates some techniques of artificial intelligence (AI) [16]. So, its closest analogue seems to be the intelligent hybrid decision support system RAO [17]. This system is designed to model the objects representable as a set of productions (IF—ELSE format). Similar to the SSM, the RAO system admits building hierarchical models and use foreign simulators. However, the RAO system has some features, which prevent its usage in the considered problems. These features are as follows: production format does not allow for significant growth of a model due to the low computer efficiency of this format; the RAO do not support convenient techniques to compare alternatives for implementation of the modeled object; the hierarchy of the model is not consistent with the structure of the object; the system is not easy to consider spatial parameters of the object. So, below we describe usage of the SSM for modeling GSRs' safety.

First, some features of taking into account various aspects of the GSRs' safety are considered and a concept of structural security is proposed to unify the tools for analyzing these aspects, then a synopsis of the main features of the SSM is given, and finally the structure of a new ontology for designing GSRs is described. It allows identifying dangerous scenarios of their operation and interactions. Finally, a methodology is given for using this ontology in designing situational digital twins to analyze the safety of GSRs creating.

2 Structural Safety of GSRs

2.1 Regulatory Framework for Maintaining Safety

The importance of ensuring the safety of various areas of human activity has led to the introduction of numerous laws regulating the rules and terminology of such activities. Apparently, the main ones should be considered Federal Laws (for example, [9, 18, 19], the requirements of which are supplemented and specified by some Government

Resolutions of the Russian Federation (in particular, [20, 21]) and the authorities of lower levels. In the further presentation, the following terms will be needed.

An *emergency situation* is a situation in a certain territory that has resulted from an accident, a dangerous natural phenomenon, a catastrophe, the spread of a disease that poses a danger to people, a natural or other disaster that may entail or has resulted in human casualties, damage to human health or the environment, significant material losses and disruption of people's living conditions [9].

A *critically important facility* (CIF) is a facility, the disruption or cessation of operation of which will lead to the loss of control of the economy of the Russian Federation, a constituent entity of the Russian Federation or an administrative-territorial unit of a constituent entity of the Russian Federation, its irreversible negative change (destruction) or a significant reduction in safety of life activity of the population [9].

A *potentially hazardous facility* (PHF) is an object on which buildings and structures of a high level of responsibility are located, or an object on which more than five thousand people can simultaneously stay [9].

Obviously, at each of the listed and other possible types of dangerous objects, emergencies of various natures can occur; accordingly, numerous types of safety are introduced in publications. The Russian-language Wikipedia article gives the following definition and lists several types of safety.

Safety is the condition in which a complex system is located, when the action of external and internal factors does not lead to processes that are considered negative in relation to this complex system in accordance with the needs, knowledge and ideas available at this stage.

Types of safety are listed as follows: aviation safety, road traffic safety, railway traffic safety, flight safety, biological safety, military safety, state safety, control systems safety, information safety, computer safety, national safety, occupational safety, fire safety, sexual safety, industrial safety, radiation safety, national safety theory, safety engineering, technosphere safety, transport safety, economic safety, environmental safety, electrical safety, energy safety, nuclear safety, military service safety, *and others.*

At first glance, it is clear that this list is not exhaustive, moreover, the types of safety listed in it depend on each other, that is, they cannot be analyzed independently.

In the corresponding English-language Wikipedia page, such a list is only implied; an example of road safety is given, for which (as, apparently, for other types of safety) it is proposed to use the following safety terms (indicators):

- *Regulatory* safety is achieved when a product or design conforms to applicable design, manufacturing, or production standards and practices, regardless of the actual safety history of the product;
- *Substantive* safety (objective safety) occurs when the actual safety history is favorable, regardless of whether standards are met or not;
- *Perceived* (subjective) safety refers to users' comfort level and perception of risk, without regard to safety standards or history. For example, traffic signals are perceived as safe ones, but under some circumstances they can increase the

number of traffic accidents at an intersection. Roundabouts generally have a good safety record, but often make drivers nervous.

Summarizing the brief analysis of the regulatory framework in the field of safety of CIFs and PHFs carried out in this subsection, it can be argued that heterogeneous and unrelated requirements are used here, there are practically no quantitative safety indicators, and it is necessary to find an integral approach to solving such problems. Proposals for unifying the analysis of tasks related to the safety of industrial facilities are given below.

2.2 *Unified Method for GSRs' Safety Analysis*

As is known (e.g., [22]), the greatest financial and material damage occurs as a result of dependent (in foreign publications—cascade or chained) failures in a conglomerate of complex systems, when any failure (initiating event) in one of them generates another failure, problems spread between systems, pile on top of each other and increase the negative consequences in geometric progression.

The most sensational examples of dependent failures were as follows:

- *Quebecevent* (1957) was a chain of numerous negative consequences of the strongest (since the beginning of the space age) geomagnetic storm that occurred during the 22nd cycle of solar activity;
- Power outage happened in the USA and Canada on August 14, 2003 (*Northeast blackout* of 2003);
- A smaller example is the spontaneous switching of traffic lights on the St. Petersburg—Murmansk railway during the magnetic storm of 2010, which fortunately did not lead to other damage.

The foregoing allows to conclude that in order to develop a unified approach to the analysis of those listed in the previous subsection and other possible aspects of safety, the only necessary and acceptable idea seems to be the construction of their mapping into a common space that allows for comparing and evaluating the degree of their influence (separately and in any combination) on the performance of a group of interrelated (one way or another) robots, identifying potentially dangerous chains of dependent failures, as well as modeling the degree of safety improvement as a result of the application of any proposed failure prevention measures. Obviously, the output of such a mapping should be a quantitative (numerical) axis of a generalized degree of danger, which can be transformed into financial indicators of damage.

Apparently, the most flexible environment for creating complex spaces today is provided within the paradigm of conceptual spaces (for example, [23, 24]), widely used in various fields for constructing taxonomies and other classification systems. Here we propose concepts and relationships that are convenient for solving the problem under consideration. Let us list some of them.

A *conceptual space* (CS) is a geometric structure composed of *object quality dimensions* (OQDs) of a certain subject area, which formalize the degree of similarity/difference of such objects in order to construct a partial order relation on a set of objects. Each OQD may include one or more physical dimensions of the space of characteristics for a certain object. Thus, the pitch parameter can be ordered on a one-dimensional scale of values from "low" to "high", while the distance in Euclidean space is modeled by a three-dimensional OQD. If an OQD is discrete one, it directly divides objects into classes. An example is the gender of an individual (though, the number of classes according to this OQD has been growing catastrophically lately, and it is increasingly difficult to establish real differences between objects of different classes).

Within a CS, each object is displayed with a certain set of *attributes* (qualities). Attribute values vary within given *domains CS*, which can be continuous or discrete. Then the CS is the Cartesian product of all domains, while any object is represented by a point in the CS, and the areas of "condensation" of objects in the CS indicate *concepts*.

Relationships of similarity/difference between objects are displayed in a CS by using the concept of *distance*; usually, multidimensional scaling [25] is used for this purpose.

The *measure of distance d* in a CS is a monotonic function $d\colon CS \times CS \to T$, where T is a completely ordered set, such that two objects are more similar, the smaller is the distance between them. This function is not necessarily linear; it can also be exponential. In the special case when the distance measure is a metric, the *betweenness relation* between objects is defined: object b is between objects a and c, if and only if $d(a; b) + d(b; c) = d(a; c)$.

To search for concepts in a CS, the following properties of its subsets are usually used:

1. A CS domain is *connected*, if it is not the union of non-empty disconnected closed sets.
2. A CS domain is *star-shaped* with respect to a point p from a certain CS region, called the *CS core*, if for any point x from CS all points between x and p belong to CS.
3. A CS domain is *convex*, if for any pair of points x and y from this domain, all points between x and y also belong to it.

Thus, the core of a given CS is the set of all its points, which have property 2.

The listed properties make it possible to divide a given CS into (significant) subspaces that contain different categories of objects; in every category, each of objects is similar to other objects of this category and differs from objects of other categories in that the values of a certain set of its attributes are in specified intervals of the domains of these attributes. The fundamental difference between such a categorization procedure (the developers called it *cognitive* (prototypical) *categorization*) from traditional categorization, where all representatives of the category are indistinguishable and equal, is that cognitive categories contain more and less typical instances (objects). The most typical representative of each cognitive category, whose properties are as close as possible to the intervals specified for this category, is called its *prototype*. Classic examples are: the prototype of the category "Russian poet" is obviously Pushkin, and the prototype of the category "fruit in Russia" (less obvious, but generally accepted) is an apple. The prototype does not always coincide with the real instance of the category; it may be an "unattainable ideal." Within a category, the degree of typicality of its instances is determined by their distance to the prototype.

The advantage of the described approach compared to the classical one is that it allows to flexibly taking into account the context of the tasks and, by setting similarity thresholds and using the Voronoy diagram (mosaic) [26] to split a CS into categories according to the nearest neighbor rule; moreover, the Euclidean metric generates convex categories.

Now we will show how to apply CSs in the problem of modeling the safety of a GSR. The main idea is to build a CS metric in the situational model of the GSR, taking into account the opinions of experts about a particular state of the modeled object, and to compare different scenarios for its functioning and interactions with other GSRs according to a general criterion for the safety of the states that form these scenarios.

To illustrate the features of the CSs in this task, it is necessary to first outline some basics of the operation of the SSM, which are necessary for further presentation. All other details and proofs of the statements made here are available in the given links to publications on this topic.

3 Structure and Main Features of the SSM

The situational modeling system [13] was developed under the guidance and with the direct participation of the author as a concretization of the ideas of situational control by D. A. Pospelov [27] and the method of structural analysis and design of complex systems (SADT, E. N. Jordan) [28] for solving various problems of modeling complex non-stationary spatially distributed objects. The core of the SSM is the *situational conceptual model* (SCM) of such object, which is open to operative changes, manages all stages of modeling and organizes the joint work of simulation models (simulators) of the object's elements with the GIS (geographical information system, electronic map) and ES (expert system) built into the SSM.

In the SCM, entities of three types are allowed, namely: *SCM objects* (organizational components of the modeled object), *processes* (simulators of the dynamics of the operation of objects) and *resources* (simulators of all interactions of the SCM objects and processes, usually in the form of time series of values of some typified variables). The development of the SCM for some system begins with the creation of a model of its organizational structure—a *tree of SCM objects*, in which the decomposition of an SCM object into sub-objects can be of two types: according to AND, when all sub-objects are necessary for the functioning of their super-object, and according to exclusive OR, when the super-object is implemented by one of its sub-objects (this is how structural alternatives to the modeled object's functioning are entered into the SCM). In each branch of the tree of SCM objects, one of them is associated with the map by assigning to it a finite number of GIS elements (for example, points, arcs, polygons) that make up its graphical representation and allow this representation to be unambiguously associated with the conceptual structure of the object under study for automatic input of the graphical characteristics of the SCM objects in simulators and for visualizing of the calculation results. Each process is assigned to an object and simulates producing a given set of its *output resources* from a known set of *input resources*. Resources can be both *material* (substances, energy, etc.) and *informational* (messages, signals, financial flows, tax rates, etc.). The former describe the real connections between the elements of the SCM, while the latter are used to configure the operating modes of the modeled system.

As a result of constructing a tree of SCM objects, assigning sets of processes to them and specifying the resources, which connect objects and processes, a model is formed as a data processing network [29]. The model can be changed as necessary by editing the SCM structure and/or specifications of its elements. After completion of the design and at any stage of modification of the model, a number of procedures are performed to control the correctness of the model that guarantees the possibility of automatic generation of simulation tasks.

The SSM belongs to applied intelligence systems since it contains an embedded expert system (ES) with a bidirectional (combined) mechanism of inference and two alternative solvers, for deterministic and probabilistic inference. Probabilistic inference is used in main operations with situations (their enriching, classifying and generalizing), while the deterministic inference provides formal correctness analysis of the SCM as well as simulation of those object's components (processes and resources) not having separate simulators. This way we provide the AI' tradition of using rapid prototypes in needed cases. Besides, the ES SSM calculates spatio-temporal functions necessary to automate processing of graphical parameters of the SCM's objects (distances, areas, etc.). In general, the SSM adheres the paradigm of semiotic formal systems [16].

Methods for implementing SCM elements in a computer are specified by assigning each of them an *executor*. This term was proposed by the author quite a long time ago in both Russian-language [30] and English-language [31] publications, however, it did not gain widespread use then, and today it has been replaced by the functionally and semantically equivalent, but much more vague term *digital twin* [32]. A more detailed discussion of this issue is given below, in Sect. 4.2.1. Any process executor specifies the rules for generating values of a set of output resources from input ones, a resource executor generates a time series of its values, and an object executor automates the issuance of some required graphical characteristics. The simulation process consists of determining the correct sequence and calling process executors to implement the required control scenario for a modeled object, automatically preparing the resource values necessary for this, and presenting the results on the map. In the absence of information for constructing mathematical models of processes and resources, their executors can be specified by a set of production rules in the SSM ES, thus building *rapid prototypes* of the corresponding models, which are necessary, at least for the study of newly created systems.

The *situational approach* to the study of the modeled object in the SSM is based on strict definitions of the following concepts, widely used in various fields of science and technology without such a justification.

A *fact* contains information about some current or desired values of a resource and is an atom of information in the SSM. An *initial situation* is a list of facts that the user enters at the beginning of a session of working with the model. This list is interpreted by the system as a simulation task. Therefore, the initial situation is expanded and enriched by analyzing the structure of the object under study (if necessary, the user's wishes are specified) to a *complete situation*, modeled by a connected fragment of the model, which contains all the data necessary to simulate the initial situation. The root object of this fragment is called the *decision-making object* (DMO), since its organizational level allows her/him to manage the complete situation. Any complete situation may contain several valid options for structuring the investigated system (they appear when constructing the model due to the decomposition of objects by OR and specifying alternative connections by resources). Within a complete situation, each alternative structure is non-redundant; it is called a *sufficient situation* in the sense that it contains enough information for imitation. Sufficient situations can be compared statically (according to special *quality criteria* assigned to each SCM object and configured by higher-level decision makers by directly setting the nominal values of the scalar quality criteria of the object and the permissible relative deviations of the current criterion values from the nominal ones) and dynamically (by simulating the behavior of the real system when input resources of this sufficient situation change according to some given sets of values).

Using the above-mentioned quality criteria for objects, which make it possible to estimate certain generalized costs of producing a particular resource of an object, the main operation of the situational control [27] is unambiguously performed, namely *classification of situations* based on the dominance of the contribution of one or another scalar criterion into the quality of the DMO. One class includes those situations where the same scalar criterion dominates (i.e., it has the minimal contribution due to its best correspondence to DM preferences); within the class, naturally, a sufficient situation and the corresponding structure of the complete situation is the better, the less are its generalized costs. Thus, in each class there is the best sufficient situation and the correspondent structure of the modeled system (we will conditionally call such a situation *optimal* one, although it is preferable only for given values of resources). At the request of the decision maker (DM), at each iteration of the simulation, the DM can leave the system in the current class of situations or change to another class, then the system offers the DM the optimal structure of the new class, and additionally takes into account the requirement of minimal system disturbances when changing the class of situations. For this purpose, the procedure for classifying situations has been advanced compared to "classical" situational control: it is performed with using the normalized Tversky model [33], which we have modified by taking into account the hierarchical structure of the modeled system.

A *scenario* in the SSM is a sequence of sufficient situations implemented during one simulation session, regardless of whether the class of situations was changed. Comparison of scenarios (for the same DMO) is performed in a *generalized modeling space*, into which variables (resources) of any type can be included, preserving the boundary conditions convenient for comparison and normalization that apply to the original generalized quality criterion for the DMO.

All studied scenarios are stored in the SSM archive and used in the future when searching for analogues to speed up the preparation of decision-making.

Now we will show how, within the framework of the SSM, the safety of a real system can be studied.

4 Detection of Failures Within the SSM

The general idea of analyzing the safety in the SSM was developed on the basis of the logical part of the logical-probabilistic approach by Ryabinin [34] and consists in the fact that each resource is associated with variables that reflect the reliability and safety of its producing under a given operation scenario, which was used for building logical models of the reliability and safety of the complex. Their significant difference between these models lies in the degree of detail: in the SCM it is increased up to a separate resource, while in the works of I.A. Ryabinin, it is based on an enlarged block diagram of the modeling object. Details of the calculations are given in [13], here we will discuss only the modifications of the conceptual modeling space necessary for this, as well as the structure of the ontology for designing *situational digital twins* (SDTs) of the SCM processes that participate in safety modeling.

4.1 CS of Safety Tasks in the SSM

The analysis of safety problems in the SSM is simplified due to the list format for storing variables of any type adopted during the development of the SCM [13]. This was done to provide a discrete approach to building models, but it also turned out to be useful when taking into account safety aspects, since the study of scenarios for the occurrence of dangerous initiating events is always associated with the work of experts [34] who are usually able to evaluate only a limited set of options for abnormal and emergency situations.

To build a CS for the task considered here, it is necessary to perform the following sequence of actions with a list of values for each potentially dangerous resource in the SCM:

- Choose its *nominal* (ideal from the point of view of the quality of the performance of the modeled object) value val_{nom}; obviously, from the point of view of cognitive classification, this value corresponds to the prototype object of a certain class;
- Arrange the remaining possible values according to the degree of their difference from the val_{nom} and place them on the CS axis on one or both sides from the ideal

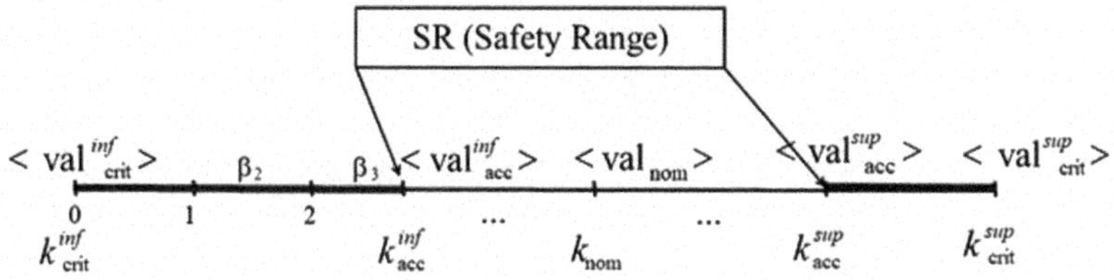

Fig. 1 Axis of resource values in the CS for safety analysis

value so that the more different values are further from the prototype, as is done on ordinal scales [35];

- Among the most different values, choose those whose implementation cannot yet lead to an emergency situation; let us call them extreme *acceptable values* (supremum and infimum, val^{sup}_{acc} and val^{inf}_{acc} in Fig. 1); there are no more than two of them: upper and lower ones; all further located values refer to dangerous values, and extreme values are *critical* (val^{sup}_{crit} and val^{inf}_{crit} in Fig. 1), at which the system collapses and an emergency situation occurs;
- Obtain expert assessments of the degree of danger of system transitions between dangerous values.

As a result of this procedure, the scale of resource values will take the form shown in Fig. 1, where integers indicate the ordinal numbers of values from left to right, symbols k with subscripts and superscripts indicate functionally significant points on the scale of values, and symbols β with indices indicate degrees of dangers for transitions between dangerous values of a variable. The resource values located between the upper and lower acceptable values constitute the safety range (SR); the remaining values are dangerous. To calculate the degree of danger of resource values, vector states of the modeled object under various scenarios, some unified dependencies were proposed [13], which preserve the conditions for normalizing the quality criterion and boundary conditions, useful for identifying the root cause of the deterioration in the quality of work of the pilot plant and the occurrence of emergency situations.

The following definitions are used.

Definition 1. Acceptable scenarios cannot include critical values of potentially dangerous resources. When such values appear, the scenario is called *prohibited* and is implemented only for simulating emergency situations.

Definition 2. A *safe scenario* can only include resource values from their safety ranges; otherwise it belongs to dangerous scenarios if it is not prohibited.

Definition 3. An *acceptable scenario* is called optimal over a given time interval if its implementation leads to a minimum of the amount of generalized costs over this interval.

Definition 4. Among dangerous scenarios, the scenario with a smaller amount of generalized costs for the development of potentially hazardous resources is preferable.

Then the goal of *strategic modeling* of a system for a given DMO on a certain time interval is to synthesize a scenario that is optimal on this interval by finding the corresponding sequence of sufficient situations with the possibility of changing the classes of situations at the choice of the decision maker. *Operational modeling* is required when some significant deterioration in the quality of functioning of the real system is detected and it is necessary to find the cause of such deterioration. For this purpose, in the tree of SCM objects, the most underlying one is identified, for which the boundary condition for its quality criterion is violated, and the operation of such an object is diagnosed. If the reason for the deterioration is not related to the increased costs of producing the input resources of this object, it is necessary to adjust the scenario of its operation by using a new classification of situations.

The fundamental difference between CSs in the SSM and traditional applications is that the dimension of this space can change significantly at each modeling step as a result of a decision made by the decision maker to transfer the modeled object from one class of situations to another and the corresponding structural changes in this object.

When dangerous scenarios appear, the presented approach allows studying the effectiveness of any planned measures to prevent abnormal and emergency situations by introducing changes in the structure of the real system, which correspond to these measures, and the acceptable scenarios generated by them.

At all stages of modeling, SDTs of SCM processes are used; some basic principles of their implementation are briefly described in the next subsection.

4.2 Ontology of SDTs Design in SSM

4.2.1 Digital Twins: Dreams and Reality

Apparently, hundreds of thousands of publications in various languages have already been devoted to this term in scientific and popular literature, so it makes no sense to provide specific links to sources. Digital twins are considered to be the best approach in solving complex problems of modeling [36]. It is expected that the topic of digital twins will remain the most popular for several years. Consequently, this has led to complete inconsistency in understanding the functionality of digital twins. General agreement has been reached only that a digital twin is a computer program, the results of which are in one sense or another comparable to the results of operating a real object. Even if we limit ourselves to technical objects, for them various sources propose to include a variety of characteristics of its prototype into the digital twin: from 3D-models to a maintenance schedule, with the ultimate goal of ideally making the digital twin a full-scale copy of the prototype, and even having an intelligent interface.

It seems that such an approach is unconstructive and unpromising to the same extent and for the same reason that the implementation of Frankenstein is unlikely. The entire history of artificial intelligence systems confirms the well-known thesis: a program works the more effectively the narrower (smaller in size) its area of competence is. Therefore, only the SDTs specialized for the task of studying the safety of complex systems are discussed below. Moreover, even narrow-profile digital twins require for specification and certification of their functional content and interfaces. The initial step in this path is taken below.

4.2.2 Functions of the SDTs Design Ontology in the SSM

Within the framework of the SSM, SDTs solve the following main tasks: to adequately display the potentially dangerous effects of a newly developed system on surrounding structures and the possible dangerous effects of the latter on the new system. As has already been stated, the most universal and flexible medium for presenting SDTs for new systems is a SDT design ontology.

Since each step of simulation modeling in the SCM solves the Cauchy problem for a certain sub-model of the SCM, this design ontology (in its analytical capabilities) should be equivalent to the state equation of the modeled system, that is, to correctly specify the dynamics of changes in its internal and output variables for given values of the input variables and the initial state of the system, including for dangerous values of state variables. The need to develop such ontology is associated with the impossibility of creating a mathematical model of complex systems, especially at the early stages of their design, without expert knowledge.

For each potentially dangerous resource of a system, the ontology must specify all the characteristic points of the scale of its values, the general view of which is shown in Fig. 1, as well as indicators of the degree of danger of transitions β_i. In the general case, these indicators depend on the values of other resources of this system and other ones related to it in some way (electrically, geographically, functionally, etc.); then the rules for calculating β_i must also be stored in the ontology. In addition, each resource or group of resources can be calculated by different executors (see Sect. 3) depending on the current situation in the modeled object, and then the ontology should contain links to the corresponding program modules of the executors and the rules for transition from one executor to another. A similar problem must be solved for executors of SCM processes: they are almost always substituted when resource values go beyond the safety range, but they can also change within this range depending on the scenario being studied.

For different complexities of the modeled system, SDT design ontologies can be created both for individual SCM processes and for groups of such processes, if the composition of these groups cannot change during the classification of situations.

If the types of output variables of the SDT design ontologies of a certain system coincide with the types of input variables of another complex or several complexes, it is necessary to simulate their interactions for different values of joint variables, especially for extreme values, which makes it possible to identify such combinations of values that can lead to abnormal and emergency situations, as well as to modify ontologies based on taking into account measures to prevent such situations.

5 Conclusion and Discussion

For preventive safety analytics during the creation of collectives of robots, a structure for the ontology of designing situational digital twins of robots within a group is proposed based on a typification of input and output potentially dangerous resources of such systems. The ontology is intended to simulate the interactions within a group or between such groups in order to quantify the degree of danger of dependent (complex, cascading) failures and the effectiveness of measures to prevent them.

The described result is in line with and at the level of global trends in the development of flexible intelligent systems for modeling complex objects being designed in order to assess their effectiveness and safety of interactions with existing objects.

In the author's opinion, the preventive safety analytics procedures presented in this work are advisable to become a stage of the feasibility study for infrastructural projects in order to reduce the possibility of dependent failures.

Meanwhile, the major features of the SSM have been tested in several main applications [13]. Every such application yielded some positive effects, but no one of them simultaneously exploited all techniques necessary to reveal dependent failures. So, one of the goals of this publication is to invite those interested in the theme to collaborate as the exhaustive proof of efficiency for the developed apparatus will take a fair time and money.

The following can be indicated as directions for further research on the topics considered:

- In addition to the Internet of Things, it is necessary to develop a new sector of the Internet, namely the Internet of Models of Things, for which to typify and standardize ontologies for designing situational digital twins of frequently encountered components of complex systems as a universal means of intellectualizing methods for solving such problems [16]. This idea is in concordance with the ideas of building the artificial intelligence of things [36];
- Provide the opportunity to purchase or rent standard data storage systems created by various developers, similar to GRID technologies or applications for cell phones. The latter option provides greater assurance of intellectual property rights for developers and allows security to be enforced when necessary.

Before implementing infrastructural projects, it is advisable to comprehensively study their safety in the virtual world of interacting models, which requires creating a flexible environment for developing this inevitable virtual world.

Acknowledgements This work was carried out within the framework of the state task for the IIMM KSC RAS (R&D "Development of Theoretical, Organisational and Technical Foundations of Information Support for Management of Viability in Regional Critical Infrastructures in the Arctic Zone of the Russian Federation" No. FMEZ-2022-0023).

References

1. Fridman, A.: SEMS-based control in locally organized hierarchical structures of robots collectives. In: Gorodetskiy, A.E., Kurbanov, V.G. (eds) Smart Electromechanical Systems: The Central Nervous System. Series: Studies in Systems, Decision and Control, vol. 95, pp. 31–50. Springer, Berlin (2017)
2. Fridman, A., Kulik, B.: Assessment of situational awareness in groups of interacting robots. In: Gorodetskiy, A.E., Tarasova, I.L. (eds) Smart Electromechanical Systems: Situational Control, pp. 25−33. Springer Nature Switzerland AG (2020)
3. Fridman, A., Kulik, B.: Task scheduling within robots' collectives of arbitrary structures. In: Gorodetskiy, A.E., Tarasova, I.L. (eds) Smart Electromechanical Systems: Behavioral Decision Making, pp. 95−100. Springer Nature Switzerland AG (2021)
4. Fridman, A., Kulik, B.: Structure of a Model of Situational Control for Intelligent Robots Containing SEMS Knowledge Bases, by means of Coordinating Contexts of a Specialized Language of Their Communication (this book)
5. Popkov, Yu.S.: Positive dynamic systems with the entropy operator. Autom. Remote. Control. **3**, 104–113 (2003)
6. Shonkwiler, R.W., Mendivil, F.: Explorations in Monte Carlo Methods. Springer, Berlin (2009)
7. Koutny, M., Robin Bergenthum, R., Ciardo, G. (eds.).: Transactions on Petri Nets and Other Models of Concurrency XVII. Springer GmbH, DE, Berlin, Germany (2024)
8. Mokayed, H., Quan, T.Z., Alkhaled, L., Sivakumar, V.: Real-time human detection and counting system using deep learning computer vision techniques. Artif. Intel. Appl. **1**(1) (2023)
9. Federal Law 68-FL dated December 21.: On the protection of the population and territories from natural and man-made emergencies (as amended on June 11, 2021). (In Russian) (1994)
10. Mesarovic, M.D., Macko, D., Takahara, Y.: Theory of Hierarchical Multilevel Systems. Academic Press, New York, London (1970)
11. Borgest, N.M.: Design ontology: genesis and development. In: Twenty-first National Conference on Artificial Intelligence with International Participation, KII-2023 (October 16–20, 2023, Smolensk, Russia). Proceedings of the conference. In 2 volumes. Vol. 1, pp. 6–13. Print-Express, Smolensk (2023). (In Russian)
12. Shvedin, B.Y.: Design ontology—TERRA INCOGNITA? Des. Ontol. **1**, 9–21 (2011). (In Russian)
13. Fridman, A.: Situational Modeling: Definitions, Awareness, Simulation. Nova Science Publishers, Inc., USA (2023)
14. Powersim. https://powersim.com/, Restored 12 Dec 2023 (2023)
15. VisSim. https://www.vissim.no/, Restored 12 Dec 2023 (2023)
16. Fridman, A.: Experience of intellectualization of situational modeling methods for discrete time-varying spatial objects. Autom. Remote Control **83**(6), 946–959. Moscow: Pleiades Publishing, Ltd. (2022)
17. Yasinovskii, S.I.: Intelligent hybrid decision support system RAO: yesterday, today, tomorrow. Myagkie Izmer. Vychisl. **12**(13), 33–50 (2018). (In Russian)
18. Federal Law 116-FL dated July 21.: On the industrial safety of hazardous production facilities (as amended on November 4, 2022) (1997). (In Russian)
19. Federal Law 119-FL dated December 29.: Town Planning Code of the Russian Federation (as amended and supplemented as of 02/03/2023) (2004). (In Russian)
20. Government Resolution 1225-GR dated 08/14/2020.: On approval of the Rules for the development of criteria for classifying objects of all forms of ownership as critically important objects (2020). (In Russian)

21. Government Resolution 304-GR dated 05/21/2007.: On the classification of natural and man-made emergency situations (as amended on December 20, 2019) (2007). (In Russian)
22. Rinaldi, S.M., Peerenboom, J.P., Kelly, T.K.: Identifying, understanding, and analyzing critical infrastructure interdependencies. IEEE Control. Syst. Mag. **21**(6), 11–25 (2001)
23. Williams, A.: Human-centric functional computing as an approach to human-like computation. Artif. Intell. Appl. **1**(2), 118–137 (2022)
24. Gärdenfors, P.: Conceptual Spaces: The Geometry of Thought. A Bradford Book. MIT Press, Cambridge, MA (2000)
25. Borg, I., Groenen, P.J.F.: Modern Multidimensional Scaling: Theory and Applications (2nd ed.), pp. 207–212. Springer, New York, (2018)
26. Gärdenfors, P.: Geometry of Meaning: Semantics Based on Conceptual Spaces. MIT Press, Cambridge (2014)
27. Pospelov, D.A. Situational Control: Theory and Practice. Batelle Memorial Institute, Columbus, OH (1986)
28. Marca, D.A., McGowan, C.L.: SADT: Structured Analysis and Design Technique. McGraw-Hill Book Co., Inc., New York, NY (1988)
29. Tyugu, E., Penjam, J.: Model-based technology of software development in large. In: Proceedings of the 14th Symposium on Programming Languages and Software Tools (SPLST'15), Tampere, Finland, October 9–10, pp. 149–163. CEUR Workshop Proceedings 1525 (2015)
30. Fridman, A.Y.: Situational approach to modeling the state of a spatial object. In: Information Support Systems for Regional Development, pp. 45–49. KSC RAS, Apatity (1998). (In Russian)
31. Fridman, A.J., Oleynik, A.G., Putilov, V.A.: GIS-based simulation system for state diagnostics of non-stationary spatial objects. In: Proceedings of the 12th European Simulation Multiconference (ESM'98), June 16–18. Vol. 1, pp. 146–150, Manchester, UK (1998)
32. Digital Twin: Transforming How We Make Sense of Data. https://www.ptc.com/ru/industry-insights/digital-twin (date of restoring 03 Nov 2022)
33. Tversky, A.: Features of similarity. Psycholog. Rev. **84**(4), 327–352 (1977)
34. Ryabinin, I.A.: Reliability and Safety of Structurally Complex Systems. Polytechnic, St. Petersburg (2000). (In Russian)
35. Bolshakov, A.A., Karimov, R.N.: Methods for Processing Multidimensional Data and Time Series. Telecom, Moscow, Hotline (2014). (In Russian)
36. Weber-Lewerenz, B.C., Traverso, M.: Navigating applied artificial intelligence (AI) in the digital era: how smart buildings and smart cities become the key to sustainability. Artif. Intel. Appl. **1**(4), 230–243 (2023)

Recognition and Classification

Acoustic Energy Mapping Method for Acousto-Optical Non-destructive Testing System with Automated Positioning

Artyom Kh. Akhmerov, Sofya D. Sycheva, Ozar I. Khodzhimukhamedov, Ivan G. Deyneka, Aleksandr S. Vasilev, and Andrey V. Kulikov

Abstract *Problem statement:* Most acoustic non-destructive pipeline testing methods operate according to one of two basic principles. The first one assumes passive long section monitoring using distributed sensors, while the second one relies on small region testing using piezoelectric transducers. However, there is another option which is to perform medium Section (0.5 to 200 m long) diagnostics by applying echo testing method using sensor arrays, distributed or quasi-distributed sensors. Even further functioning improvement can be achieved by acoustic components positioning automation implemented with SEMS modules. One of the main problems with this approach is the complexity of reconstruction and visualization of spatial information about defects locations. *Purpose of research*: Formation of the acoustic energy mapping method for the intelligent acousto-optical non-destructive pipeline testing system with automated positioning that allows echo-signal source map visualization to reveal locations of defects and inhomogeneities. *Results:* The work proposes an architecture for the pipeline testing system with sensors and emitters installed on a SEMS-based pipe crawler robot. It uses magnetic chassis for movement and verifies current position using rangefinder data. The paper presents

A. Kh. Akhmerov (✉) · S. D. Sycheva · O. I. Khodzhimukhamedov · I. G. Deyneka · A. S. Vasilev · A. V. Kulikov
ITMO University, Saint-Petersburg, Russia
e-mail: akakhmerov@itmo.ru

S. D. Sycheva
e-mail: sdbochkova@itmo.ru

O. I. Khodzhimukhamedov
e-mail: oihodzhimukhamedov@itmo.ru

I. G. Deyneka
e-mail: igdeyneka@itmo.ru

A. S. Vasilev
e-mail: a_s_vasilev@itmo.ru

A. V. Kulikov
e-mail: avkulikov@itmo.ru

I. L. Tarasova and B. A. Kulik (eds.), *Smart Electromechanical Systems*, Studies in Systems, Decision and Control 544, https://doi.org/10.1007/978-3-031-64277-7_9

a computer model that simulates sound propagation through the metal pipe material inside an area with a size of 500 × 500 mm. The model adds defects with size 0.5 to 50 mm of various shapes to random locations. Then it applies the acoustic energy mapping method that performs defect location visualization. During modelling the method skipped about ~3.6% defects. The defect location measurement error (computed as object centroid displacement) did not exceed 5.2 mm. *Practical significance:* The results of the work form the technical and algorithmic basis for pipeline testing systems of a new class designed for a medium-length pipeline section diagnostics.

Keywords Acousto-optical sensing · Non-destructive testing · Pipeline testing · Ultrasonic testing · Acoustic energy mapping · Beamforming · Acoustic imaging

1 Introduction

Ultrasonic methods of pipeline non-destructive testing play an important role in ensuring the safety and reliability infrastructure of the fuel and energy complex of several countries [1–6]. They provide the possibility to detect and locate defects, assess the degree of their danger, measure wall thickness and others. This allows to timely and promptly prevent possible emergencies and to reduce the rate of their occurrence. It is a well-known fact that acoustic non-destructive testing systems involve two main measurement modes: diagnostics and continuous monitoring. One of the modern examples of an acoustic continuous monitoring system is distributed fiber-optic systems known as DAS (Distributed Acoustic Sensing) [7–10]. Such systems offer significant advantages including an ability to continuously monitor the condition of a sufficiently long section of the pipeline, as well as the defect development dynamic [11]. In addition, if sensitive optical fiber is built inside the measured object, an additional opportunity opens up to protect strategically important sections of pipelines and prevent illegal "tapping" for the purpose of stealing resources. The most significant disadvantages of such systems are the low accuracy and sensitivity of measurement, as well as the cost of their acquisition and operation.

Examples of diagnostic systems for acoustic non-destructive testing are ultrasonic flaw detectors and thickness gauges. They are actively used in practice, as well as their modifications containing point fiber-optic sensors [12–14]. An important positive feature of such systems is the high accuracy of recording the acoustic signal. The key disadvantage is the limited measurement area. At the same time, some of the researches [12, 15] are aimed at creating quasi-distributed acoustic non-destructive testing systems. This makes it possible to overcome this limitation by increasing the number of sensitive elements. To summarize, the use of acoustic monitoring methods using passive distributed or an array of point optical-acoustic sensors allows for effective diagnostics of oil and gas pipelines [7–14]. However, the key difficulty of working with such systems is the need to restore spatial information based on reference signal sources and sensitive elements locations, as well as large volumes of recorded data and their subsequent processing.

We propose to use an acoustic energy mapping method for metallic pipeline testing that covers medium length sections (up to 200 m). This approach allows to visualize the acoustic map of wave reflection sources, which reveals the locations of several inhomogeneities or defects in the area being tested. Proposed method is designed to operate in conjunction with a non-destructive testing system for metal pipelines with automated positioning of ultrasound generators and sensitive elements. *The main purpose of the research* is to develop the acoustic energy mapping method for the acousto-optical system for pipeline non-destructive testing. Its core function is to visualize an echo signal source map to identify locations of defects and inhomogeneities.

2 Acousto-Optical Non-destructive Testing System Design

There is a large variety of different solutions for device positioning automation, especially for steel objects. One of the most successful solutions uses magnetic chassis that allow robotized systems to stick to either internal or external surfaces of the pipes [16]. With this approach, some commercial devices already can cover significant sections of pipelines, up to 180 m [18–20]. Moreover, it is suitable for most commonly used pipeline diameters, from 100 to 1000 mm and more.

To solve pipeline testing problems including area limitations and spatial information reconstruction complexity we propose our solution for acousto-optical pipeline testing system design. The core of the system includes fiber optic sensors array combined with multiple ultrasonic sources. The intelligent system is designed for ultrasonic pipeline testing that combines multiple features to minimize human participation in inspection process. It concerns hardware peculiarities as well as algorithms and software. And naturally, its architecture allows to provide decent level of performance to compete with existing solutions. Figure 1 shows the proposed architecture.

The proposed architecture of the non-destructive testing system for pipelines contains four acoustic signal sources and four sensors installed on a SEMS-based robot that ensures movement and positioning along the central axis of the pipeline. Reference acoustic signal works in pulse mode inducing diverging longitudinal waves that propagate through the pipe material. The rangefinder improves positioning precision by measuring the absolute coordinate relative to the pipeline to fix cumulative error. Acoustic components perform echo-testing on the adjacent pipeline section, receiving the echo-signal from every reflectance source. Signal processing block preprocesses all the signals acquired with acousto-optical sensors. Once all the data is acquired, the specialized software applies acoustic energy mapping algorithm to visualize defect locations in accordance with base positioning data. The defect location map reconstruction can be implemented either inside signal processing block, or at the remote operator device.

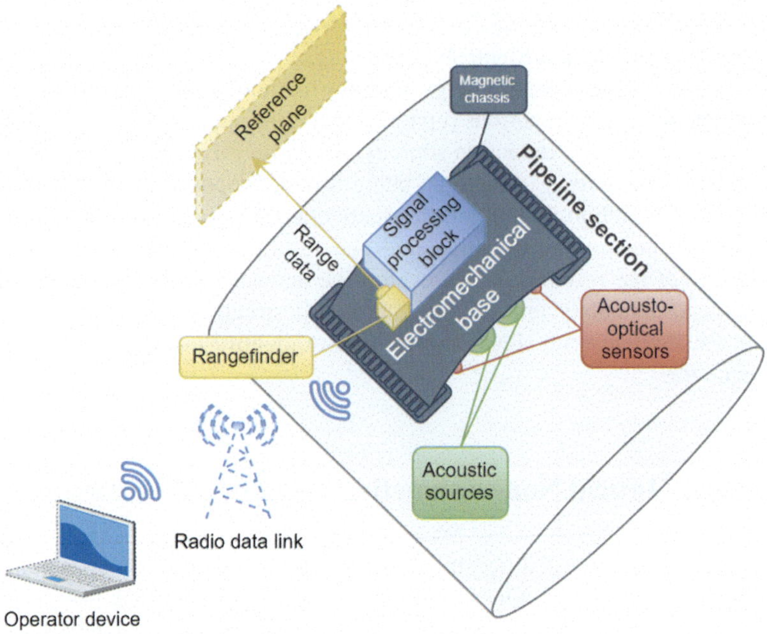

Fig. 1 Intelligent acousto-optical pipeline testing system architecture

The main requirement for system acousto-optical components is to allow echo-testing to cover the whole cross-section of a pipe in a single measurement. In this case, the spatial interval between measurement stops is defined mostly by reference acoustic source power and acousto-optical sensors sensitivity. With this requirement met, the systems positioning task can be simplified to the movement along the top of the pipeline with cross-sectional angular coordinate preservation. Both velocity and acceleration data can define current cross-sectional angular coordinate, and magnetic chassis controller can keep it constant.

The four-sources and four-sensors configuration allows to use beamforming techniques for ultrasonic pulse-echo testing implementation. Although the pulse-echo testing method is widely used for various tasks, it requires very high sampling frequency to provide reasonable defect mapping resolution. This frequency level is achievable by using contemporary acoustic fiber-optical sensors (i.e. Fabry–Perot interferometers, Bragg grating sensors) and powerful signal processing devices, such as FPGAs (field programmable gate arrays). The most energy-effective and compact option for acoustic source implementation is based on piezoelectric transducers, so it is the best choice to use for autonomous solutions.

3 The Acoustic Energy Mapping Method

We propose the acoustic energy mapping method that relies on beamforming techniques and applies a portion of algorithms from the field of image processing to visualize the map of defect locations.

(1) The first step is to perform acoustic echo-signal measurements. The electromechanical base firmly maintains the relative positions of each ultrasound source and each acoustic sensor. So, we can use signal arrival time for each sensor to reconstruct the spatial information about acoustic wave reflector locations. Each ultrasound source produces signal sine wave pulse $P_g(t)$. The pulses from different sources are separated with attenuation time t_a that is required to wait for medium stress relaxation. Pulse start time $t_{g,0}$ for each source g can be calculated using following formula:

$$t_{g,0} = \left(t_p + t_a\right) \cdot g, \tag{1}$$

where $t_{g,0}$ is the pulse start time for generator at index g (starting with zero), t_p is the pulse duration, t_a is the attenuation interval to wait for stress relaxation. The acoustic pressure signal is defined on the finite interval of duration t_p and starting time $t_{g,0}$:

$$P_g(t) = \begin{cases} m_g \cdot \sin\left(2\pi f_g \cdot \left(t - t_{g,0}\right) + \varphi_0\right) \cdot \left(1 - \frac{t - t_{g,0}}{t_p}\right), t_{g,0} \le t \le \left(t_{g,0} + t_p\right) \\ 0, \left(t < t_{g,0}\right) \vee \left(t > \left(t_{g,0} + t_p\right)\right) \end{cases}, \tag{2}$$

where $P_g(t)$ is the acoustic signal value at time t, m_g is the acoustic source signal magnitude, f_g is ultrasound frequency, φ_0 is the initial sine signal phase. Each sensor acquires 4 data sequences, each corresponding to one particular ultrasound source position. According to that, the acoustic part forms 16 digital signal sequences $P_{g,s}(t_k)$ for output, where g is the sound generator index, s is sensor index, and t_k is data sample time coordinate at index k.

(2) The visualization starts with a canvas, and in this case its spatial resolution depends on the sampling frequency and user requirements and is limited by following value:

$$R_{\min} = \frac{F_s}{c}, \tag{3}$$

where F_s is the acoustic sensor sampling frequency, and c is the average speed of sound wave propagation through the pipe material. The thickness of pipe walls is significantly lower than the diameter. Due to this fact, we can simplify the visualization to two-coordinate representation, as if sound waves were propagating through the flat object. This simplification reduces the software development complexity and saves a great amount of computational resources with minimal performance reduction. Once the resolution is defined, we can prepare the canvas, that represents a two-dimensional digital signal *I*(*X*, *Y*) of size *M* × *N* samples. *X* is longitudinal section coordinate index (*0 to M* – *1*) and *Y* is cross-sectional coordinate index (*0 to N* – *1*). The next equation describes the correspondences between visualization point coordinates (*X*, *Y*) and spatial coordinates (*x*, *y*):

$$\begin{cases} x = X \cdot R \\ y = Y \cdot R \end{cases}, \tag{4}$$

where *X* and *Y* are the visualization point indexes, *x* and *y* are spatial coordinates measured in meters, *R* is the resolution measured in meters per unit. The *Y* coordinate value can be easily (but not necessarily) converted to angular coordinate which is commonly-used for pipeline testing.

(3) Time of echo-signal arrival depends on the ranges between reference signal source, potential reflectance source location and acousto-optical sensor. The next step is to calculate the time of arrival for each canvas point and each generator-sensor pair:

$$t_{g,s}(x, y) = \frac{\sqrt{\left(x - x_g\right)^2 + \left(y - y_g\right)^2} + \sqrt{(x_s - x)^2 + (y_s - y)^2}}{c}, \tag{5}$$

where $t_{g,s}(x, y)$ is the time of arrival between ultrasound generator at coordinates (x_g, y_g) and acoustic sensor at coordinates (x_s, y_s) through the point at coordinates (*x*, *y*).

(4) Time of arrival in case when wave propagates directly from source to sensor is less than if it goes through any reflecting point. According to that, we can cut out the signal parts that correspond to direct reception from reference source. This step is required because of the considerable magnitude difference between direct impact and echo.

(5) The reflectance signal shape partially corresponds to the reference signal. This fact allows to filter non-corresponding acoustic perturbations out. The valid echo-signal value from point (*x*, *y*) is defined as a covariance between acoustic source pulse shape and sensor acoustic data considering propagation time delay and attenuation weight function:

$$I_{g,s}(x, y) = \omega\left(t_{g,s}(x, y)\right) \cdot \frac{1}{n_p} \sum_{k=0}^{k=n_p-1} P_{g,s}\left(t_k + t_{g,s}(x, y)\right) \cdot P_s(t_k) - \overline{P_{g,s}} \cdot \overline{P_s}, \tag{6}$$

where $I_{g,s}(x, y)$ is the echo-signal intensity from point at coordinates (*x*, *y*) acquired from the reference ultrasound source *g* with sensor *s*, t_k is the acoustic signal sample acquisition time coordinate at index *k*, T_s is the acoustic signal sampling period, n_p is the reference ultrasound pulse duration measured in sampling periods T_s. $\overline{P}_s$ and $\overline{P}_{g,s}$ are the source and sensor acoustic pressure signals respectively. $\omega(t_{g,s}(x, y))$ is the weight function that allows to take attenuation into account during intensity value calculation:

$$\omega\left(t_{g,s}(x,y)\right) = \frac{1}{e^{-2\cdot\delta\cdot t_{g,s}(x,y)\cdot c}}, \tag{7}$$

where δ is the pipe material attenuation coefficient. Time $t + t_{g,s}(x, y)$ in Eq. (5) does not correspond to acoustic signal sampling time coordinates. To avoid that, we can calculate $P_{g,s}(t)$ signal interpolation as a weighted average of two nearest signal samples:

$$P_{g,s}(t) = P_{g,s}(t_1)\cdot\left(1 - \frac{|t - t_1|}{T_s}\right) + P_{g,s}(t_2)\cdot\left(1 - \frac{|t - t_2|}{T_s}\right), \tag{8}$$

where $P_{g,s}(t)$ is the interpolated signal value at time t, t_1 and t_2 are two closest sample time coordinates to the time t, T_s is the acoustic signal sampling period.

(6) The next step is to combine echo-signal intensity values acquired from each acoustic sensor into a single intensity map. This step allows to verify each acoustic reflector location through echo-signal acquisition check by using different times of arrival for each sensor. We propose a simple but efficient way of combination:

$$I_g(x,y) = \prod_s I_{g,s}(x,y), \tag{9}$$

where $I_g(x, y)$ is the echo signal intensity map produced by source g in conjunction with all sensors.

(7) From this step and further we will use point indexes (X, Y) instead of spatial coordinates (x, y) to describe discrete coordinate space processing operations. The connection between them is designated in Eq. (4). Defect representations on intensity maps that were acquired on the previous step might have a low contrast. Contrast enhancement can be achieved by applying gamma correction:

$$I_{g,e}(X,Y) = I_g(X,Y)^{\gamma}, \tag{10}$$

where $I_{g,e}(X, Y)$ is the enhanced version of intensity map $I_g(X, Y)$, γ is correction coefficient. In case of current method, most of the times the task for this transformation is to partially equalize the contrast of dim and bright objects. Therefore, coefficient γ is supposed to be less than 1 for almost any implementation of this algorithm.

(8) Objects visualized on intensity maps $I_{g,e}(X, Y)$ might be loose due to the way of spatial information reconstruction. We apply the grayscale morphological closing [20] to make intensity map objects more solid:

$$I_{g,\cdot}(X, Y) = I_{g,e}(X, Y) \cdot S = \left(I_{g,e}(X, Y) \oplus S\right) \ominus \mathrm{S}, \tag{11}$$

where $I_{g,cl}(X, Y)$ is the result of morphological closing, $I_{g,e}(X, Y)$ is the source intensity map, $\cdot$ denotes the morphological closing operator, $\oplus$ and $\ominus$ denote the morphological dilation and erosion operators respectively, S is the structuring element. Structuring element indicates a subset of coordinates in the proximity of each two-dimensional signal point to calculate the new brightness value. The morphological dilation and erosion operations can be performed using following equations:

$$I_{g,\ominus}(X, Y) = \min\left(I_{g,e}(A_S(X, Y))\right), \tag{12}$$

$$I_{g,\ominus}(X, Y) = \min\left(I_{g,e}(A_S(X, Y))\right), \tag{13}$$

where $I_{g,\oplus}(X, Y)$ and $I_{g,\ominus}(X, Y)$ are the dilation and erosion results respectively, $A_S(X, Y)$ is the subset of two-dimensional signal point coordinates in proximity of point (X, Y) according to the structuring element S. The preferable shape of the structuring element is a disk, because it most closely matches the wave front. The size of structuring element is mostly defined by wavelength of the reference acoustic pulse and its value must exceed $\lambda \div R$.

(9) The main point of two previous steps is the preparation for intensity map fusion operations [21]. And the main purpose of this step is the further verification of defect locations. To achieve it, we perform another intensity map multiplication similar to the Eq. (9):

$$I(X, Y) = \prod_{g} I_g(X, Y), \tag{14}$$

where $I(X, Y)$ is the resulting acoustic energy map. The two-dimensional signal constructed at this step can be further used for defect localization and classification. Defect localization task can also be automatically solved by means from the field of image processing, or performed manually by operator. Same goes for classification task which solutions can rely either on convolutional neural networks or on human perception.

4 Computer Modeling and Results

We present a computer model that consists of three main parts. The first part implements acoustic wave propagation simulation in the medium with defects of various shapes at random locations. The second part implements the proposed acoustic energy mapping method and provides defect location visualization. The third part performs visualization location error assessment using morphological processing and initial simulation mask. The model applies Matlab [22] platform along with k-Wave

Toolbox [23] expansion for sound propagation simulation. The model parameters are as follows: propagation medium is steel with sound speed $c = 5800$ m/s $\pm$ 1% and medium density $\rho = 7800$ kg/m^3 $\pm$ 1%, reference source signal is represented by a sine wave with frequency 100 kHz and pulse duration $t_p = 50$ µs, acoustic sensors sampling frequency $F_s = 5$ MHz.

The model composes a simulation mask that sets the positions of acoustic sources and sensors, as well as defect locations. Sensor and source coordinates are constant throughout the simulations. At the same time, defect locations were randomized for each simulation and their number varied from 1 to 5. They are composed from multiple sequential line segments. Each of line segments has length from 0.1 to 5 mm and the number of segments for one defect varies from 1 to 20. However, total length is restricted to 50 mm. Model applies additional shading to each defect to increase its roughness. The defects on the mask are considered as cracks filled with air with sound speed $c_d = 335$ m/s and density $\rho_d = 1.27$ kg/m^3. The whole computer model processing pipeline is shown at Fig. 2.

The *first stage* that simulates reference acoustic signal propagation through steel considering all the main medium properties and wave interactions. The model carries out a simulation for each ultrasonic source position gathering acoustic pressure data

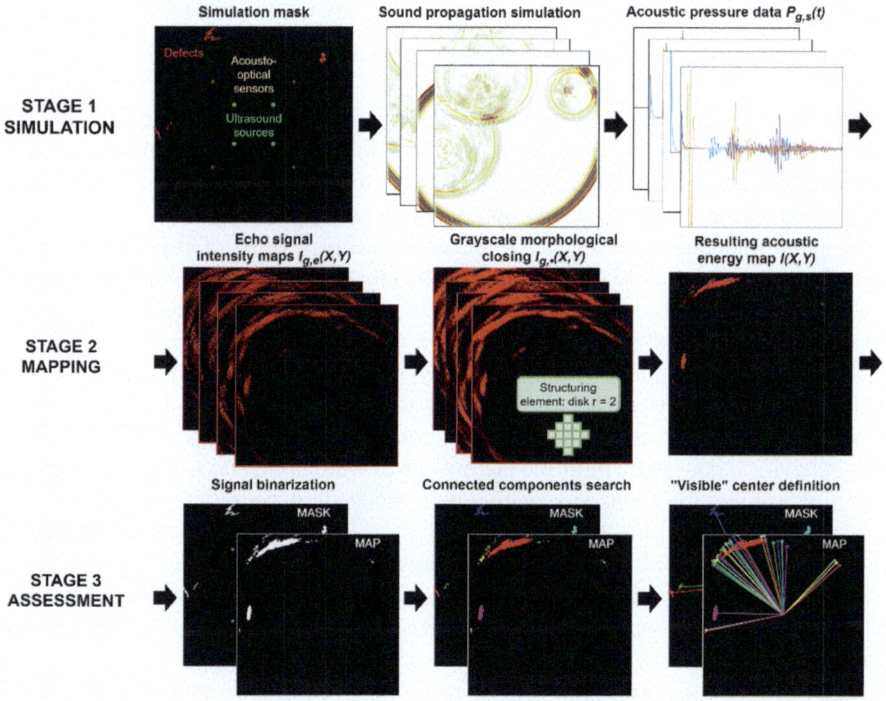

Fig. 2 Acoustic energy mapping method computer model structure

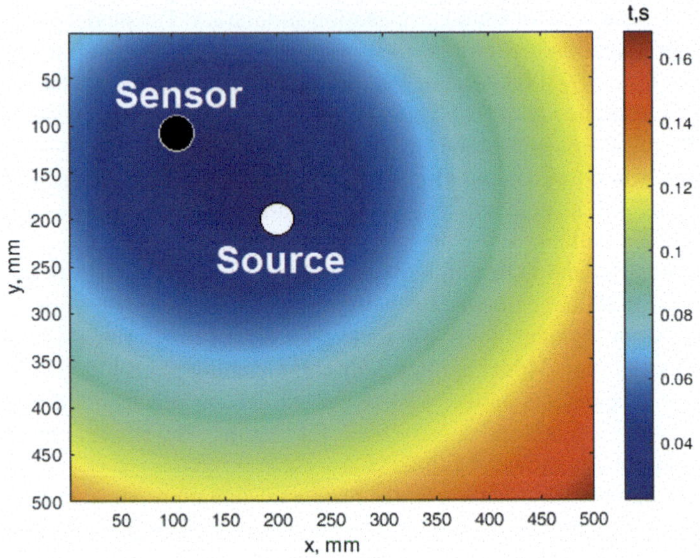

Fig. 3 Propagation time distribution

$P_{g,s}(t)$. Data collection is performed using all four sensors in accordance with acousto-optical system configuration.

The *second stage* performs the acoustic energy mapping method. It computes all time of arrival $t_{g,s}(x, y)$ values according to Eq. 5. Figure 3 shows the elliptical sectors that characterize propagation time increasement for particular generator and sensor locations. After arrival time calculation the model transforms acoustic data $P_{g,s}(t)$ to echo signal intensity maps $I_{g,e}(X, Y)$ according to Eq. (6–10). Grayscale morphological closing (Eq. 11–13) makes all the objects on the intensity maps more solid. Following intensity map fusion operation (Eq. 14) produces the resulting acoustic energy map that can be used for defect location determination.

The *third stage* applies adaptive energy map binarization and separates it to multiple sets of points each corresponding to a single object (connected component). However, the shape of defects does not correspond to the initial real shapes. This behavior is caused by multiple factors one of which is the *invisibility* of the most part of the object for acoustic sensitive part. Acoustic wave reflection occurs once it reaches the closest object on its way. If there is another object behind the closest one it will not be visualized because the acoustic waves did not even reach it. Same goes for longitudinal size (along the wave propagation direction) of *visible* objects. According to this factor, the longitudinal size of object on the acoustic energy map is not valid.

To determine the correct centroid of each object we find the closest object point that lies on the line between average sound generator position and conventional object centroid. The centroid coordinates are calculated as average values of object point set coordinates. Similarly, source average coordinate is calculated. The resulting centroid is determined as an object point that lies on the line between acoustic source and object centroids, that is the closest one to sources location. We choose the point (X_c, Y_c) for which there are two conditions that must be met. Firstly, the point must be within a line between sources centroid and object centroid. Secondly, the range between sources centroid and object point must be minimal. To check first condition, we calculate the range between an object point (X, Y) and source-to-object line using following formula:

$$r_l(X, Y) = \frac{\left|(Y_{c,o} - Y_{c,g}) \cdot X - (X_{c,o} - X_{c,g}) \cdot Y + X_{c,o} \cdot Y_{c,g} - Y_{c,o} \cdot X_{c,g}\right|}{\sqrt{(Y_{c,o} - Y_{c,g})^2 + (X_{c,o} - X_{c,g})^2}}, \quad (15)$$

where $r_l(X, Y)$ is the range between an object point (X, Y) and source-to-object line, $(X_{c,o}, Y_{c,o})$ are the coordinates of object centroid, $(X_{c,g}, Y_{c,g})$ are the average coordinates of sound generator positions. Then, to check second condition we calculate the range between assumed centroid (X_c, Y_c) and ultrasound generator location $(X_{c,g}, Y_{c,g})$:

$$r_g(X, Y) = \sqrt{(X - X_{c,g})^2 + (Y - Y_{c,g})^2}, \quad (16)$$

where r_g is the range between point (X, Y) and ultrasound generator location $(X_{c,g}, Y_{c,g})$. Following equation represents these two conditions to determine the resulting visible object centroid:

$$(X_c, Y_c) \in (r_l(X_c, Y_c) < 1) \wedge \left(r_g(X_c, Y_c) \rightarrow \min\right), \quad (17)$$

where (X_c, Y_c) are the resulting visible object centroid coordinates. The model iterates through all of the object points checking these conditions to determine its visible centroid coordinates. It performs this procedure both for defect mask as well as for resulting acoustic energy map.

Right after that, the model determines the defect localization error as a range between each mask object centroid and the closest resulting acoustic energy map object centroid:

$$E = \sqrt{(X_{c,r} - X_{c,m})^2 + (Y_{c,r} - Y_{c,m})^2}, \quad (18)$$

Here E is the defect localization error, $(X_{c,r}, Y_{c,r})$ are the object centroid coordinates on the resulting acoustic energy map. $(X_{c,m}, Y_{c,m})$ are the object centroid coordinates on the defect mask.

We carried out 500 simulations using presented computer model. In total, the model placed 1590 defects (3.18 per simulation). During model functioning, 57 objects were not visualized either due to the very small size or specific positioning that minimized the reflecting border length. It corresponds to a skipping probability

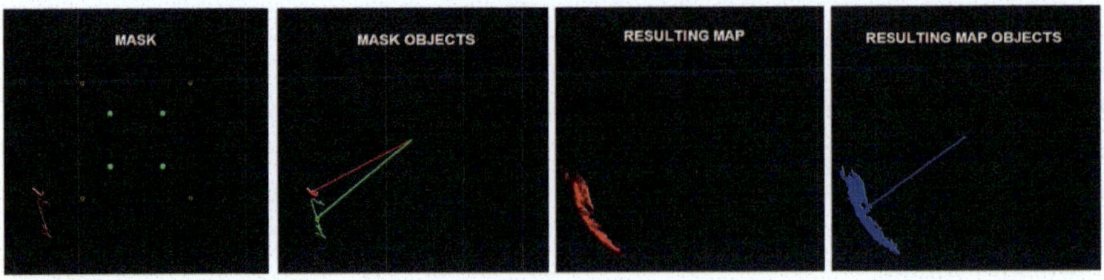

Fig. 4 Object visualization merging case example

around ~3.6%. Another problem is acoustic energy map object merging. Figure 4 illustrates the example of the merging case.

There were 31 problematic cases in which one resulting map object was composed of two separate echo-signal sources with a very short range between them. For those cases mask centroid was recalculated as an average centroid between those objects to match it with the resulting acoustic energy map. Figure 5 presents the dependency of the error estimation results on the distance between acoustic map object centroid and ultrasound generator locations.

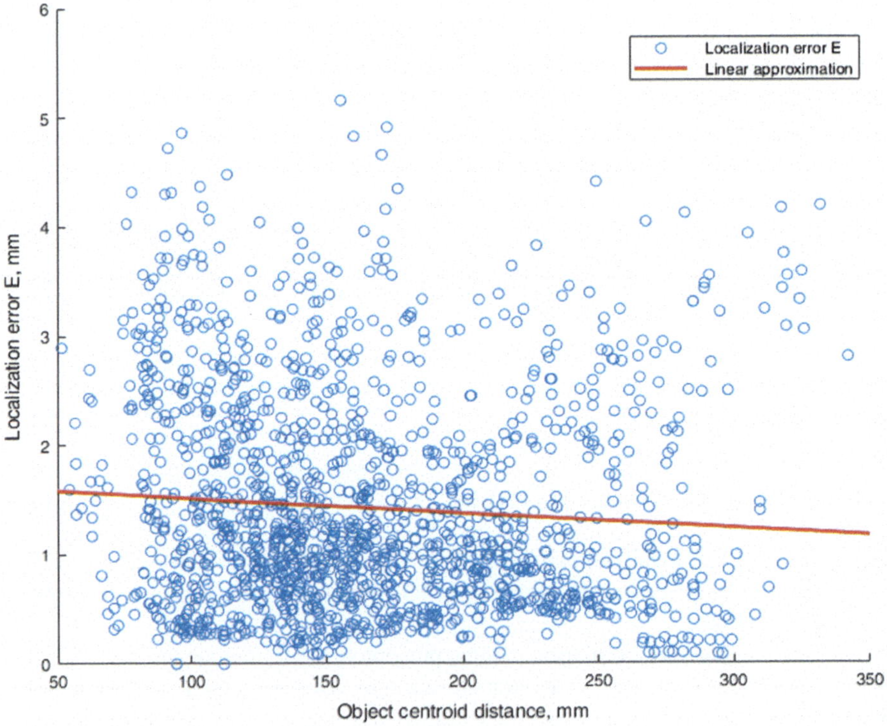

Fig. 5 Localization error estimation results

The figure shows that localization error slightly decreases as the range from sound generators to object increases. Objects that are close to sound generators reflect waves that come from different source positions differently. Indeed, the waves that come from different sides collide with different object borders with significantly distinguishing coordinate values. This fact affects the localization accuracy especially for objects that are close to acoustic signal generators. According to the calculations, the mean localization error value is $\overline{E} = 1.49$ mm and the standard deviation is $\sigma = 1.02$ mm.

5 Conclusion

The work proposes an architecture for the pipeline testing system with sensors and emitters installed on a SEMS-based pipe crawler robot. It uses magnetic chassis for movement and verifies current position using rangefinder data. It uses radio data link to transfer all the acquired data for further processing. All the acoustic, acousto-optical, electronic and opto-electronic components are installed on the electromechanical base. These features allow the acousto-optical non-destructive testing system to achieve high degree of autonomy.

The article proposes the acoustic energy mapping method for acoustic echo-testing. The method reconstructs spatial data using the acoustic pressure signal from four sensors and four ultrasound generators. It relies on echo signal time of arrival calculations and estimates the echo-signal intensity as its covariation with sound generator signal shape. After initial intensity map reconstruction, the method significantly enhances it with contrast transformations and grayscale morphological closing. The acoustic energy mapping method still has significant flaws that we must get rid of. The most crucial among them is the high number of false objects on the acoustic energy map that increase false alarm probability.

The paper presents a computer model that simulates sound propagation through the metal pipe material inside an area with a size of 500×500 mm. The model includes three stages: sound propagation simulation, acoustic energy mapping and defect localization error estimation. The first stage adds 1 to 5 defects with size of 0.5 to 50 mm of various shapes to random locations and simulates sound propagation through the steel pipe section. The second stage applies proposed acoustic energy mapping method that performs defect location visualization. The third stage finishes modelling by calculating acoustic energy map object centroids and estimating localization error. During the simulation the localization error did not exceed 5.2 mm, mean error value is $\overline{E} = 1.49$ mm and the standard deviation is $\sigma = 1.02$ mm.

As a conclusion, the proposed acoustic energy mapping method proved to be highly efficient at determining steel object defect locations. The results of the work form the technical and algorithmic basis for pipeline diagnostic systems of a new class designed for a medium-length pipeline section testing.

Acknowledgements The present work was financially supported by grant RSF No. 23-79-01274.

References

1. Zang, X., Xu, Z.-D., Lu, H., Zhu, C., Zhang, Z.: Ultrasonic guided wave techniques and applications in pipeline defect detection: a review. Int. J. Press Vessels Pip **206**, 105033 (2023), ISSN 0308-0161. https://doi.org/10.1016/j.ijpvp.2023.105033
2. Stüwe, I., Zacherl, L., Grosse, C.U.: Ultrasonic and impact-echo testing for the detection of scaling in geothermal pipelines. J. Nondestruct. Eval.Nondestruct. Eval. **42**, 18 (2023). https://doi.org/10.1007/s10921-023-00926-0
3. Santos, D., Machado, M.A., Monteiro, J., Sousa, J.P., Proença, C.S., Crivellaro, F.S., Rosado, L.S., Santos, T.G.: Non-destructive inspection of high temperature piping combining ultrasound and eddy current testing. Sensors **23**(6), 3348 (2023). https://doi.org/10.3390/s23063348
4. Suslov, A.V., Yaroslavkina, E.E.: Acoustic methods for measuring the thickness of deposits on the inner wall of pipelines, Petrol. Eng. **21**(5), 132–144 (2023). ISSN 2073-0128, https://doi.org/10.17122/ngdelo-2023-5-132-144
5. Rai, A., Ahmad, Z., Hasan, M.J., Kim, J.-M.: A novel pipeline leak detection technique based on acoustic emission features and two-sample kolmogorov-smirnov test. Sensors **21**(24), 8247 (2021). https://doi.org/10.3390/s21248247
6. Evgeny, B., Dumitrescu, A., Parinov, I.: Non-destructive testing and repair of pipelines (2018). https://doi.org/10.1007/978-3-319-56579-8
7. Wu, Y., Gao, L., Chai, J., Li, Z., Ma, C., Qiu, F., Yuan, Q., Zhang, D.: Overview of health-monitoring technology for long-distance transportation pipeline and progress in DAS technology application. Sensors **24**(2), 413 (2024). https://doi.org/10.3390/s24020413
8. Bertulessi, M., Bignami, D.F., Boschini, I., Longoni, M., Menduni, G., Morosi, J.: Experimental investigations of distributed fiber optic sensors for water pipeline monitoring. Sensors **23**(13), 6205 (2023). https://doi.org/10.3390/s23136205
9. Baensch, F., et al.: Feasibility study: continuous monitoring of pipes using distributed acoustic and fiber optic sensors. In: International Symposium on Structural Health Monitoring and Nondestructive Testing, 4–5 Oct 2018, Saarbruecken, Germany (2018)
10. Tanimola, F., Hill, D.: Distributed fibre optic sensors for pipeline protection. J. Nat. Gas Sci. Eng. **1**(4–5), 134–143 (2009). ISSN 1875-5100. https://doi.org/10.1016/j.jngse.2009.08.002
11. Gorshkov, B.G., Yüksel, K., Fotiadi, A.A., Wuilpart, M., Korobko, D.A., Zhirnov, A.A., Stepanov, K.V., Turov, A.T., Konstantinov, Y.A., Lobach, I.A.: Scientific applications of distributed acoustic sensing: state-of-the-art review and perspective. Sensors (Basel) **22**(3), 1033 (2022). https://doi.org/10.3390/s22031033.PMID:35161779;PMCID:PMC8838753
12. Rose, J.L., Philtron, J., Liu, G., Zhu, Y., Han, M.: A hybrid ultrasonic guided wave-fiber optic system for flaw detection in pipe. Appl. Sci. **8**(5), 727 (2018). https://doi.org/10.3390/app8050727
13. Moradi, H., Parvin, P., Ojaghloo, A., Shahi, F.: Ultrasensitive fiber optic Fabry Pérot acoustic sensor using phase detection, Measurement **172**, 108953 (2021). ISSN 0263-2241, https://doi.org/10.1016/j.measurement.2020.108953
14. Wu, Q., Okabe, Y., Yu, F.: Ultrasonic structural health monitoring using fiber bragg grating. Sensors (Basel). 2018 Oct 11; **18**(10), 3395 (2018). PMID: 30314260; PMCID: PMC6209894. https://doi.org/10.3390/s18103395
15. Guo, H., Xiao, G., Mrad, N., Yao, J.: Fiber optic sensors for structural health monitoring of air platforms. Sensors **11**(4), 3687–3705 (2011). https://doi.org/10.3390/s11040368
16. Kumar, S., Mahto, D.: Recent trends in industrial and other engineering applications of non destructive testing: a review. Int. J. Sci. Eng. Res. **4**(9), (2013), Sept 2013, Available at SSRN: https://ssrn.com/abstract=2770922
17. Ismail I.N. et al.: Development of in-pipe inspection robot: a review. In: 2012 IEEE Conference on Sustainable Utilization and Development in Engineering and Technology (STUDENT), pp. 310–315. Kuala Lumpur, Malaysia (2012). https://doi.org/10.1109/STUDENT.2012.6408425
18. Using crawler technology for deteting pipeline defects: nexxis, 24 Jan 2020. https://nexxis.com/using-crawler-technology-for-detecting-pipeline-defects/

19. Pshenin, V., et al.: Robot crawler for surveying pipelines and metal structures of complex spatial configuration. Infrastructures **7**(6), 75 (2022)
20. Woods, R.E., Gonzalez, R.C.: Digital image processing, 4th Edition. Pearson 20 Mar 2017, p. 1192
21. Akhmerov, A.K., Vasilev, A.S., Vasileva, A.V.: Research of spatial alignment techniques for multimodal image fusion. In: Proceeding SPIE 11059, Multimodal Sensing: Technologies and Applications, 1105916 (21 June 2019); https://doi.org/10.1117/12.2526030
22. The MathWorks Inc.: MATLAB version: 9.13.0 (R2022b). Natick, Massachusetts: The MathWorks Inc. (2022). https://www.mathworks.com
23. Treeby, B.E., Cox k-Wave, B.T.: MATLAB toolbox for the simulation and reconstruction of photoacoustic wave-fields. J. Biomed. Opt. 15(2), 021314 (2010)

Analysis of the Degree of Differentiation Biological Tissues for Diagnosis of Pathological Processes Using Robotic Spectral Complexes

Ruslan D. Khlynov, Victoria A. Ryzhova, Ekaterina A. Tsygankova, Igor A. Konyakhin, Valery V. Korotaev, Sergey N. Yarishev, Todor S. Djamiykov, and Marin B. Marinov

Abstract *The subject of the research*: The work examines the spectral characteristics of biological samples of human internal organs taken from patients diagnosed with adenocarcinoma at different stages and localizations, in particular in the primary focus of occurrence (stomach) and with subsequent metastasis to distant organs. The possibility of conducting a robotic analysis and drawing conclusions about the state of pathological biological tissue according to the degree of their differentiation based on the identification of the dependence of the measurement of the spectral characteristics of fluorophores on the stage of pathology development is considered. *Purpose of research*: Determination and analysis of the influence of the degree of differentiation on the spectral characteristics of fluorescence of tissues of human internal organs. *Results*: Spectral patterns of measurement of luminescence excitation are obtained. The quantitative dependence of the values of the extremes of the main

R. D. Khlynov (✉) · V. A. Ryzhova · E. A. Tsygankova · I. A. Konyakhin · V. V. Korotaev · S. N. Yarishev
ITMO University, Saint-Petersburg 197101, Russia
e-mail: ruslan384ruslan3@gmail.com

V. A. Ryzhova
e-mail: victoria_ryz@itmo.ru

I. A. Konyakhin
e-mail: iakoniakhin@itmo.ru

V. V. Korotaev
e-mail: vvkorotaev@itmo.ru

S. N. Yarishev
e-mail: sniaryshev@itmo.ru

T. S. Djamiykov · M. B. Marinov
Technical University of Sofia, Sofia 1756, Bulgaria
e-mail: tsd@tu-sofia.bg

M. B. Marinov
e-mail: mbm@tu-sofia.bg

I. L. Tarasova and B. A. Kulik (eds.), *Smart Electromechanical Systems*, Studies in Systems, Decision and Control 544, https://doi.org/10.1007/978-3-031-64277-7_10

fluorophores on the degree of malignancy of the pathological formation was revealed. A feature has been noted regarding the degree of differentiation of tumors. The difference in the main maxima at the excitation wavelengths of 286–290 nm and 330–340 nm increases with increasing differentiation. Thus, the peak difference in a highly differentiated tumor is approximately 3.5 times, moderately differentiated–2.5 times, and low differentiated – 1.5 times. With total organ damage by pathology, there is practically no difference in extremes. The prospects of using the research results in the development and operation of robotic spectral complexes based on SEMS modules in medicine to improve the effectiveness of diagnosing the degree of pathological changes in the body are substantiated. It is shown that the use of the obtained quantitative estimates in the robotic spectral analysis of pathological tissues of internal organs will eliminate the influence of a subjective factor on the decision on the development of pathology, more accurately determine the degree of development of the pathological process and identify its spread to neighboring healthy tissues. *Practical significance*: The results obtained can be used to develop robotic spectral complexes that will allow for more accurate and timely diagnosis of various diseases, including cancer, as well as analyze the spectral characteristics of tissues and identify abnormalities associated with pathologies.

Keywords Biological sample · Spectral characteristics · Fluorescence · Fluorophore · Differentiation · Luminescence · Laser · Robotic spectral complex · Pathology · Tumor

1 Introduction

Technologies for noninvasive diagnosis and treatment of surgical pathologies using affordable and non-traumatic methods with sufficient sensitivity and reliability are becoming increasingly in demand all over the world. Histological examination of internal organs based on non-contact tissue biopsy remains the most common, due to the convenience of the procedure [1, 2].

In medicine, the possibilities of obtaining diagnostic information based on biopsy results using optical and optoelectronic methods of tissue examination, often combined by the term "optical biopsy", are expanding. The advantage of these methods lies in the possibility of conducting research and obtaining information in real time without the need to obtain biological material from the body, as with a conventional biopsy [3, 4].

Optical biopsy techniques can be divided into optical spectroscopy and imaging techniques. Most of the optical analysis methods are occupied by a group of spectral research methods. The availability of a large amount of detailed information allows us to consider and analyze the qualitative and quantitative parameters of materials. Among the optical spectroscopic methods of in vivo diagnostics, the methods of fluorescence spectroscopy are currently the most developed. They are based on the effect of optical radiation of a certain wavelength on the tissue under study and the registration of fluorescence spectra excited by it. Fluorescence spectroscopy is an actively used highly sensitive method for studying the structure and properties of cells, as well as the biochemical reactions occurring in them. This method allows recording and analyzing in vivo the content of substances in tissues and organs that determine the metabolism and viability of cells and tissues [5–7].

One of the important applications of fluorescence spectroscopy is to assess the state of metabolic activity of tissues by studying the fluorescence intensity of NADH (nicotinamide adenine dinucleotide) and FAD (flavinadenine dinucleotide) coenzymes, which are important components of the Krebs cycle. NAD and FAD coenzymes are organic molecules that bind to enzymes and are involved in cell biological processes such as redox reactions and in the catalysis of certain metabolic reactions [8].

The study of the fluorescence intensity of NAD and FAD coenzymes makes it possible to assess the activity of redox processes in the cell, since the level of fluorescence depends on the concentration of these coenzymes and their oxidative or reduced state. Due to the fact that the optical properties of these fluorophores are sensitive to the metabolic status of biological tissue, the measurement of NADH and FAD fluorescence can be used to assess the functioning of cellular metabolism, diagnose diseases and study the biochemical pathways of cellular reactions. Therefore, fluorescence spectroscopy is a valuable tool for studying the state of biological tissues (healthy, inflamed, malignant) [9].

The fluorescence spectrum of phenylalanine has a maximum at a wavelength of 282 nm, and its quantum yield is still an order of magnitude lower than that of tyrosine. The high sensitivity of the considered fluorophores to the effects of optical radiation is actively used in determining the localization of foci of oncological pathologies using laser-induced fluorescence spectroscopy, which is one of the directions of development of medical and biological sciences.

Despite the achievements of modern endoscopy and successes in the early diagnosis of neoplasms, stomach cancer is the third most fatal cancer in the world. Currently, preclinical and clinical studies show that the use of diagnostic methods based on the registration of fluorescent radiation can significantly increase the effectiveness of early diagnosis of tumors of various localization, and therefore accelerate the treatment process [10–12].

For more accurate and objective diagnosis of tumors and early detection of pathological processes, it is necessary to use more advanced technologies, such as robotic spectral complexes. The use of Smart Electromechanical Systems (SEMS) in this context will automate the process of obtaining and analyzing spectral information, providing more accurate and reliable diagnostic results. These specialized systems combine high accuracy and efficiency, which makes them an ideal choice for improving the visualization and analysis of pathological changes in biological tissues.

Despite the potential and prospects of robotic spectral complexes based on SEMS, it should be noted that this area has not yet been fully developed. For the successful implementation and implementation of such technologies in medical practice, it is necessary to conduct preliminary studies aimed at optimizing the functionality, improving the accuracy and reliability of these systems, as well as adapting to the specifics of the diagnosis of pathological processes in biological tissues.

Objective determination of the localization of an oncological tumor and its metastases in other organs is one of the directions of development of medical and biological sciences. During a person's lifetime, including with the progression of a malignant tumor in his organs, many different molecular processes occur in biological tissues that directly affect spectral characteristics. The study of the fluorescence of excited particles allows us to obtain information about the state of living systems without damaging them, and does not require a large amount of biological material, which is very important, taking into account the aspect of working with biological tissues in vivo.

The study of fluorescence spectra obtained from histological sections of benign and malignant neoplasms of a number of organs allows us to identify features associated with the histoarchitectonics of the organ and various pathogenesis of its conditions. However, the use of spectral methods to verify metastases (MTS) of gastric adenocarcinoma (AK) is currently not widespread enough in the early diagnosis of the disease. In this regard, the development and research of new methods of differentiation of pathological gastric tissues based on the analysis of fluorescence spectra is relevant. The aim of the work is to investigate the possibility of using the fluorescence spectroscopy method to verify the degree of development of the pathological process, specifically adenocarcinoma.

To recognize the stage of development of pathological processes in medicine, the concept of "differentiation" is used. The higher the degree of differentiation, the less malignant the tumor is.

In oncology, the international index of the degree of differentiation of a tumor focus is usually designated by the letter G "grate" [13]. The signs by which tumor differentiation can be performed are shown in Table 1.

In this paper, the influence of the degree of tumor differentiation, namely, the similarity of the cellular composition of the tumor with the original cells of the organ, on the type of fluorescence spectra is considered.

Table 1 Characteristic features for tumor differentiation

Degree of differentiation	Features
G1	The tumor is located in one layer and does not grow beyond it
G2	Limited hearth up to 2 cm (for some organs, the size may be up to 5 cm)
G3	The tumor has the same size, but it has affected the lymph node closest to the focus
G4	A tumor of any size that has at least one metastasis in a distant organ or grows into neighboring organs

2 Methods and Materials

The measurements were carried out on the "Fluorat-02" spectroscopic apparatus. Its optical scheme is shown in Fig. 1.

The analyzer's light source, a high-pressure xenon lamp, operates in short pulse mode ($\approx 1\ \mu s$), with a repetition rate of 25 Hz. The emission spectrum of a xenon lamp ranges from hard ultraviolet (190 nm) to near infrared (2.5 μm) radiation. Most fluorescent substances have a narrow-band structure of absorption spectra in the wavelength range from 200 to 600 nm. He was chosen as the job.

To isolate the required spectral range, the analyzer uses a monochromator with a concave diffraction grating operating in the first order of diffraction. In order to prevent the penetration of second-order diffraction radiation into the cuvette with the analyzed sample, the monochromator is equipped with a device that cuts off the second order of diffraction. It usually turns on at about a wavelength of 400 nm. The measurements were carried out at a temperature of 27 °C, since the radiation of the samples in the area of room temperatures is of the greatest practical interest.

Light from a pulsed radiation source passes through an excitation monochromator, which highlights the spectral region of excitation, and enters a beam-splitting plate, which divides the luminous flux into two channels: a reference and luminescence excitation, which is simultaneously a photometric channel.

In the reference channel, the radiation, bypassing the sample, enters the radiation receiver of this channel, forming an electrical comparison signal, which serves to correct the instability of the lamp from pulse to pulse.

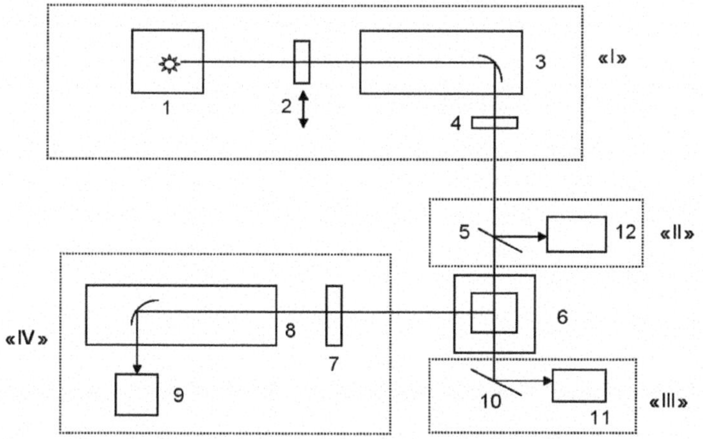

Fig. 1 Optical scheme of the "Fluorat-02-Panorama" spectrofluorimetric analyzer: 1 is a radiation source; 2 is a second-order diffraction cut-off device; 3 is a monochromator of the lighting channel; 4 and 7 are light filters of the channels of excitation and registration of luminescence; 5 and 10 are beam-splitting plates; 6 is cuvettes with the analyzed object; 8 is a monochromator of the fluorimetric channel; 9 is a photodetector of the fluorimetric channel; 11 is a photodetector of the channel transmission; 12 is photodetector of the reference channel

In the luminescence excitation channel, light passes through the test sample, causing its luminescence, and then enters its photodetector. The electrical signal of this receiver depends on the directional transmission coefficient of the object under study.

In the luminescence registration channel, the radiation of the luminescent components of the object under study passes through the registration monochromator, which highlights the spectral registration region, and enters the radiation receiver of the luminescence registration channel. The electrical signal of this receiver depends on the concentration and composition of the substances being determined in the sample.

With the help of a microprocessor system of analyzers, signals from photodetectors of all channels are processed and the concentration of the substances to be determined is calculated.

The material for the study in this work were tissue samples of such internal organs as the stomach, lymph node and omentum. The objects were presented in the form of tissue plates – thin slices of a piece of organ imprinted in medical paraffin for safekeeping. In order for the paraffin to not give an additional glow, it was previously removed from the samples by means of a xylene solution. Thus, the potential error factor was removed.

In this work, studies of highly differentiated (G1), moderately differentiated (G2), low differentiated (G3) and totally affected (G4) objects were carried out. Ten objects were obtained for the study and each object was assigned its own degree of pathology–adenocarcinoma in advance: two samples are highly differentiated, three moderately differentiated, four low differentiated and one with total affect.

For each measured sample, graphs of the spectral scanning results and the values of the wavelengths corresponding to the highest excitation index η were obtained. The ratio of these maxima to each other was the basis for further reasoning.

A research methodology has been developed, which includes the following stages:

Stage I. Preparing the installation and samples for operation

Before use, the spectrofluorometer should be warmed up for a couple of minutes, and after that download the software "Panorama PRO" and wait for it to synchronize with the device. Since the samples have been pre-waxed in xylene solution, excess liquid should be allowed to drain before starting measurements. To do this, the samples are placed in filter paper for a few minutes. After that, each sample prepared for measurement is placed alternately on the sensitive surface of the receiver of the device.

Stage II. Measurement of the excitation spectrum during synchronous scanning

At this stage, the measurement of the synchronous excitation spectrum makes it possible to identify the effect of radiation of which wavelength leads to the most intense luminescence of the studied biological object.

Preliminary settings are made (the type of scanning, wavelength, registration range, number of flashes, sensitivity of the photoelectronic multiplier, pitch are indicated) and the first measurement of the synchronous spectrum is performed in the widest possible spectral range–the device allows you to obtain correct data in the wavelength range 210–840 nm. It is advisable to set the sensitivity of the photoelectronic multiplier low at this stage of measurement in order to exclude the possibility of overshooting in the registration channel. The number of flashes of the radiation source should be chosen small (10 or 25 pcs.), and the adjustment step of the monochromator is 2 nm. The offset should be set to 70 nm, and then it should be clarified after preliminary measurement. The true value of the offset can be determined only after measuring the luminescence and excitation spectra by finding the distance between the maxima of these spectra. Usually this distance is in the range of 60–100 nm. Having received a preliminary result and estimated the approximate location of the working area, you can proceed to a clarifying measurement of the synchronous spectrum. In this case, the spectral range should be narrowed to the boundaries of the working area, and the number of flashes should be increased to 50–100. This will allow for the most accurate measurement of the synchronous excitation spectrum. Peak values and parameters such as intensity and position along the wavelength axis are searched.

Stage III. Measurement of the luminescence spectrum(s)

When measuring luminescence (scanning by registration), it is necessary for the excitation monochromator to set a constant wavelength at which the maximum of the synchronous spectrum is observed. The spectral range for the registration monochromator should be set as follows: the left boundary of the range should be set at least 20 nm longer than the wavelength of the excitation monochromator; the right boundary of the range should be set at least 20 nm less than twice the wavelength of the excitation monochromator.

If there are several maxima of the synchronous spectrum, then a registration scan should be performed for each of these maxima. The luminescence spectrum of the object under study is also analyzed – the search for the maximum.

Stage IV. Measurement of the excitation spectrum

During this scan, a constant wavelength must be set for the registration monochromator, and the spectral range and scanning step must be set for the excitation monochromator. The registration monochromator should be installed at the same wavelength at which the maximum of the luminescence spectrum is observed. The spectral range for the excitation monochromator should be set as follows: the right boundary of the range should be set at least 20 nm less than the wavelength of the registration monochromator; the left edge of the range must be set at least 20 nm more than half the wavelength of the registration monochromator. During the trial scan, the number of flashes of the radiation source is small, the step of adjustment of the excitation monochromator is the largest, the sensitivity of the photoelectronic multiplier is the same as with synchronous scanning. To search for peaks as an analysis of the excitation spectrum.

Stage V. Construction of excitation and luminescence spectra and their general analysis

The distance between the maxima of these spectra is equal to the actual value of the offset during synchronous scanning. You can also perform mathematical operations with graphs (calculation of logarithms and exponents, calculation of derivatives of different orders and integrals, smoothing of various types) after which, the results from the "Panorama" program are transferred to others, for example, "Excel", where they are analyzed and modified in the future.

Stage VI. Error estimation

The evaluation criteria are established by probability theory and error theory. Therefore, the accuracy is estimated based on random errors. The order of finding the relative error in determining the maximum intensity value in the spectrum during n multiple direct measurements:

(1) Calculate the arithmetic mean of the measured value using the formula:

$$\overline{x} = \frac{x_1 + \cdots + x_i + \cdots + x_n}{n}. \tag{1}$$

(2) Calculate the estimate of the mean square deviation using the formula:

$$S_{\overline{x}} = \sqrt{\frac{\sum_{i=1}^{n} (x_i - \overline{x})^2}{n \cdot (n-1)}}. \tag{2}$$

(3) Calculate the confidence interval of the random error (random error) using the formula:

$$\Delta_{\overline{x}} = t_{\alpha,n} \cdot S_{\overline{x}}, \tag{3}$$

where $t_{\alpha,n}$ is the Student's coefficient, which takes into account the required confidence probability α and the quantitatively performed measurements n, on the basis of which the value of the mean square deviation $S_{\overline{x}}$ is calculated.

(4) Determine the absolute measurement error taking into account the random error $\Delta_{\overline{x}}$ and the instrumental error Δ_u according to the formula:

$$\Delta_x = \sqrt{\Delta_{\overline{x}}^2 + \left(\frac{2}{3} \cdot \Delta_u\right)^2}, \tag{4}$$

where the 2/3 multiplier takes into account the different confidence probabilities of determining a random $\Delta_{\overline{x}}$ and instrumental Δ_u of errors. The random error is calculated for a confidence probability $\alpha = 0.95$, and the value of the instrumental error Δ_u of the device is normalized for a confidence probability of $\alpha = 1$.

(5) Calculate the relative measurement error using the formula:

$$\varepsilon_x = \frac{\Delta_x}{\overline{x}} \cdot 100\%. \tag{5}$$

Each type of scanning (synchronous, luminescence, excitation) was performed sevenfold for one sample. After that, the values were averaged and a single picture of each measurement was obtained, the main points of attention in which were extremes at certain wavelengths. Thus, the error for the wavelength $\lambda = 336$ nm was calculated: $n = 7$; $\overline{x} = 9.52$; $S_{\overline{x}} = 3.66$; $\Delta_{\overline{x}} = 6.95$; $\Delta_x = 2.19$; $\varepsilon_x = 23.05\%$.

Similarly, the error calculation was carried out for the remaining wavelength parameters. So, $\varepsilon_{\lambda 286} = 23.47\%$; $\varepsilon_{\lambda 480} = 23.7\%$. Based on this, it can be concluded that the results of this study have a sufficient degree of approximation to the actual value of the value.

3 Discussion of the Results

Highly differentiated cells cause forms of cancer that grow slowly, rarely go beyond the organ in which they were formed, and metastasize much more slowly. Consequently, their spectral patterns will be similar to tissues without pathologies. Fig. 2 shows the spectrum of a highly differentiated gastric adenocarcinoma sample. Several pronounced peaks are noticeable at the excitation wavelengths of 262, 270, 286 nm, as well as 330 and 418 nm single ones, most likely related to the organ itself, and not to the site with a pathological change.

The spectrum of this gastric adenocarcinoma sample demonstrates the strongest excitation of luminescence at wavelength 286 nm (Fig. 3).

In sample 6436-37/16, metastases to the lymph node (L/N) also have many extremes (Fig. 4), due to the specifics of the structure of the organ tissue. This is a reticular (connective tissue), in the loops of which lymphocytes are located, which give a characteristic reaction to the action of radiation.

The main extremes at $\lambda_{\text{excitation}} = 230$, 250, 260 and 286 nm, as well as in the long wave region at 336 and 426 nm (Fig. 5).

It is worth noting that the tissue of the lymph node itself is practically absent.

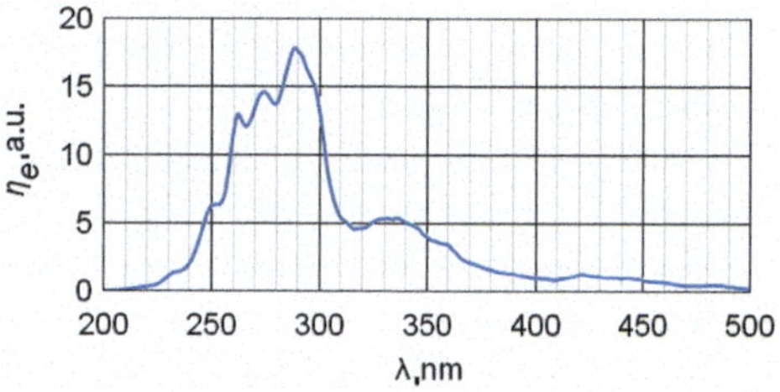

Fig. 2 Scanning spectrum of the 6447/16 gastric adenocarcinoma sample with a high degree of differentiation G1

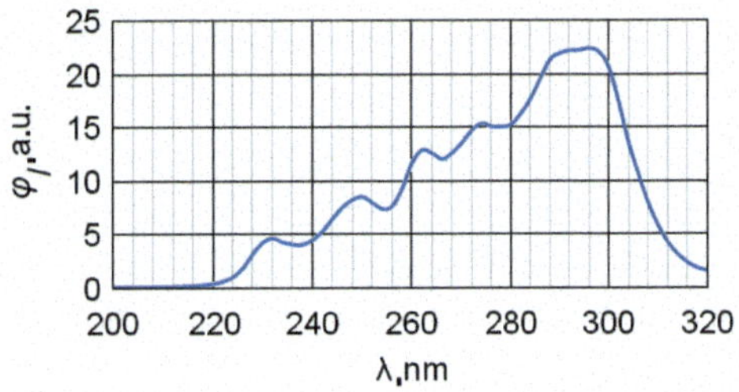

Fig. 3 The luminescence scanning spectrum of the 6447/16 gastric adenocarcinoma sample with a high degree of differentiation G1 by $\lambda_{excitation} = 286$ nm

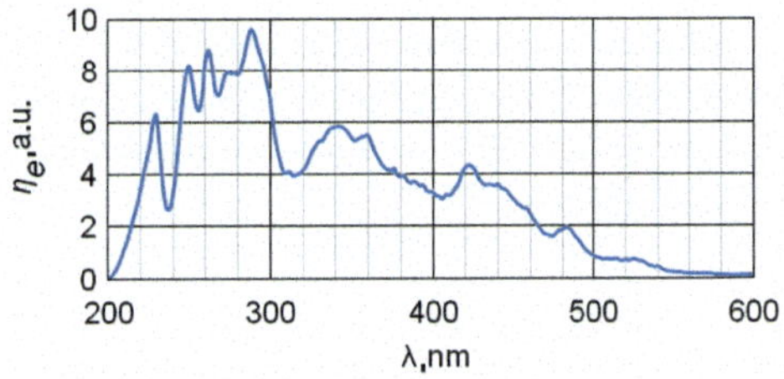

Fig. 4 Scanning spectrum of the 6436–37/16 of lymph node metastasis sample with a high degree of differentiation G1

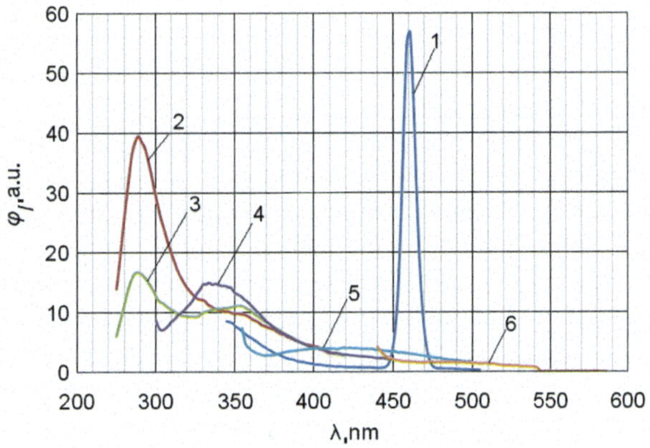

Fig. 5 Luminescence spectra of sample 6436-37/16 of lymph node metastasis 6436-37/16 with a high degree of differentiation G1: 1 is at $\lambda_{excitation} = 230$ nm; 2 is at $\lambda_{excitation} = 262$ nm; 3 is at $\lambda_{excitation} = 250$ nm; 4 is at $\lambda_{excitation} = 288$ nm; 5 is at $\lambda_{excitation} = 342$ nm; 6 is at $\lambda_{excitation} = 424$ nm

Moderately differentiated adenocarcinoma is a transitional form in which there are more tumor cells and they spread more aggressively. Moderately differentiated gastric adenocarcinoma samples (Fig. 6) are qualitatively similar, since they have the same number of maxima at approximately equal wavelengths.

All the samples are united by the fact that the ratio between the peaks of short waves is about 2 times.

Samples of omentum metastases from different patients were examined, which showed mixed results, shown in Fig. 7. They differ from each other in peak ratios.

As can be seen, the difference in extremes for sample (**b**) is about 2.25 times, while for (**a**) sample it is 1.2 times.

Upon further examination, it turned out that the sample (**a**) contains fat and connective tissue overgrowth. It is known that the absence or small amount of the protein

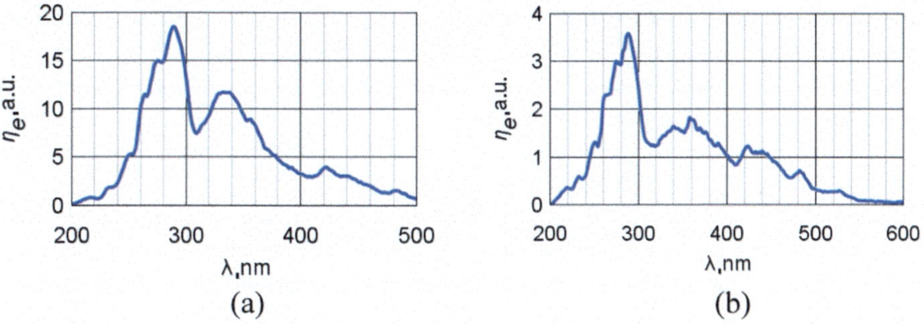

Fig. 6 Scanning spectra of gastric adenocarcinoma samples with a moderate degree of differentiation G2: **a** 9183/16; **b** 10132/16

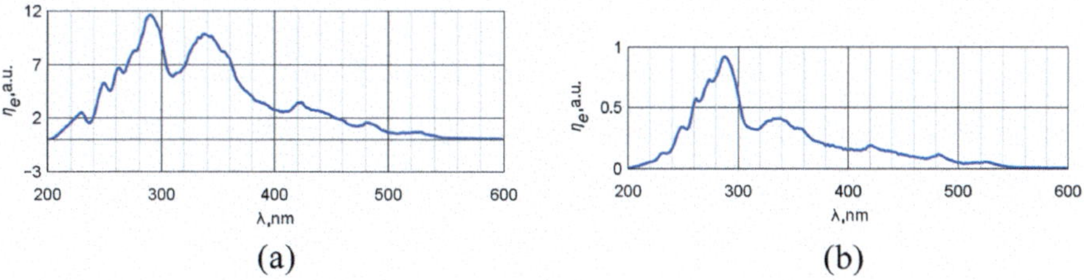

Fig. 7 The spectrum of synchronous scanning of samples of metastases in the omentum of the stomach with a moderate degree of differentiation G2: **a** 9273/16; **b** 10132/16

component contributes to the production of "Cptla" inhibitor and osteopontin, which contribute to the proliferation and metastasis of cancer cells. Based on this, we can draw the following conclusion: despite the fact that both samples were declared to be moderately differentiated, it can be said from the spectral patterns that the sample (**a**) is rather low-differentiated.

The main luminescence extremes for sample 10132/16 correspond to the excitation wavelengths $\lambda_{excitation} = 288$ and 336 nm (Fig. 8).

Low-grade adenocarcinoma is the most dangerous form of tumor, in which cells quickly degenerate into malignant ones, losing their morphological structure, and the

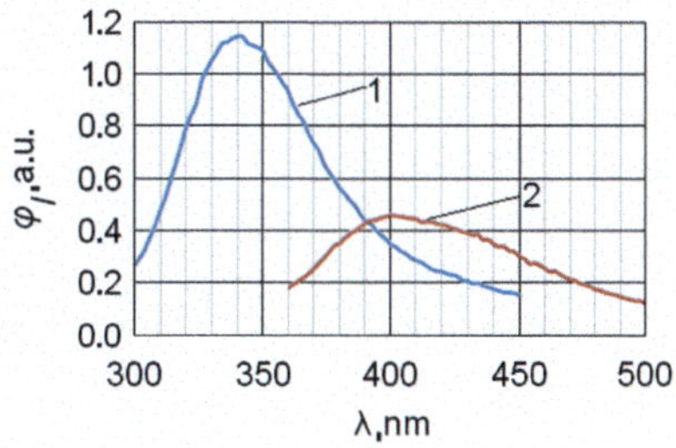

Fig. 8 Luminescence spectra of sample 10132/16 metastasis to the omentum with a moderate degree of differentiation G2: 1 is at $\lambda_{excitation} = 288$ nm; 2 is at $\lambda_{excitation} = 336$ nm

rapid spread of the tumor to other tissues begins. These include samples of gastric adenocarcinoma, the spectra of which are shown in Fig. 9.

All of them are qualitatively similar to each other and have a quantitative difference in the ratio of peaks of about 1.2–1.5 times.

The main luminescence extremes for sample 1811/16 correspond to the excitation wavelengths $\lambda_{excitation}$ = 290 and 338 nm (Fig. 10).

There degree of differentiation of G4 indicates that there is no possibility of treatment, since the cells behave extremely aggressively and finally lose the features of normal ones. The total defeat of cricoid cancer has a sample of 25568/16 gastric tissue, the spectrum of which is shown in Fig. 11.

It differs from the previous ones in that the peaks at $\lambda_{excitation}$ 290 nm and 330 nm are almost equal in height.

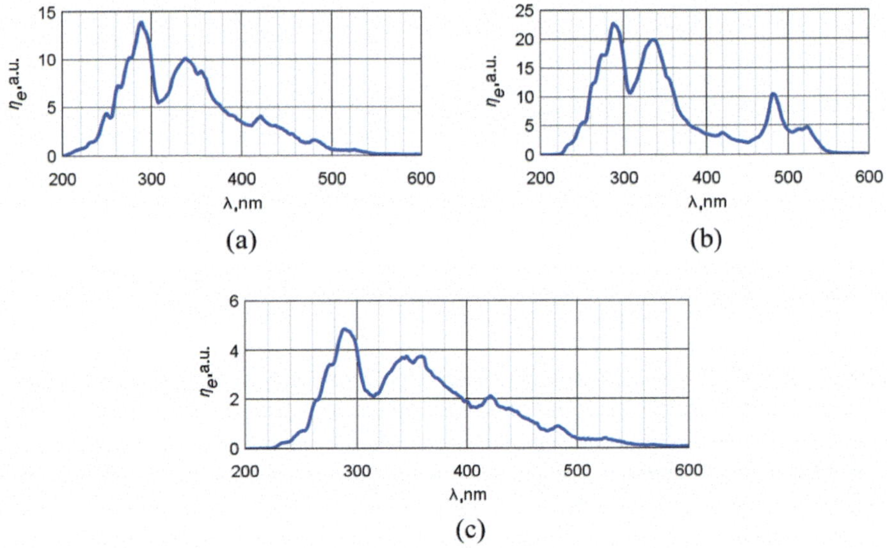

Fig. 9 Scanning spectra of gastric adenocarcinoma samples with a moderate degree of differentiation G3: **a** 1811/16; **b** 2765/16; **c** 207/19

Fig. 10 Luminescence spectra of gastric adenocarcinoma sample 1811/16 with a moderate degree of differentiation G3: 1 is at $\lambda_{excitation}$ = 290 nm; 2 is at $\lambda_{excitation}$ = 338 nm

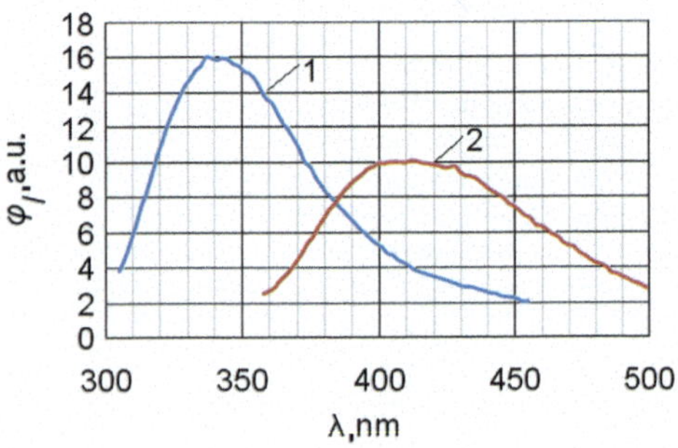

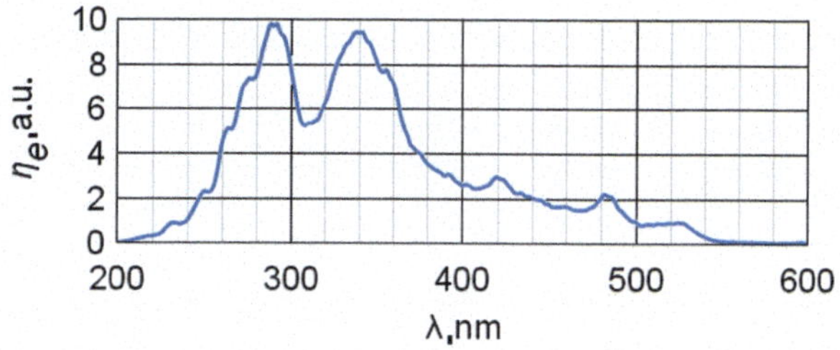

Fig. 11 The scanning spectrum of the stomach sample 25568/16 with a degree of differentiation G4

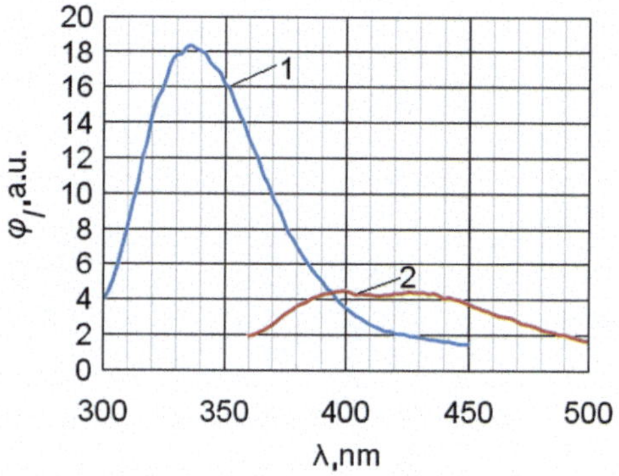

Fig. 12 Luminescence spectra of sample 25568/16 with a degree of differentiation G4: 1 is at $\lambda_{excitation} = 288$ nm; 2 is at $\lambda_{excitation} = 336$ nm

The main luminescence extremes for sample 25568/16 correspond to the excitation wavelengths $\lambda_{excitation} = 288$ and 336 nm (Fig. 12).

4 Conclusion

The resulting spectra are influenced by a number of factors, such as the state and type of changes in tissues, and most importantly, the phase development of pathological processes. But it is worth noting that there are other factors such as blood filling of the tissue, the presence of other fluids, etc. We did not take them into account in this study. Accordingly, there is a need to conduct research to study the contribution of each factor to tissue fluorescence in order to correctly interpret diagnostic data. In addition, it is impossible to talk about the correctness of the results obtained without comparing them with the fluorescence spectra of tissues of the studied organs in the absence of pathology.

According to the ratio of the main extremes, it is possible to identify the degree of differentiation of the tumor. The results are presented in Table 2.

As a percentage for each level of pathological lesion, the difference between the main peaks is: G1 is more than 60%; G2 is from 30 to 50%; G3 is from 10 to 20%; G4 is less than 10%.

Table 2 The results of the study of the spectra of tissue samples with pathological formations of varying degrees

The area under study	Sample number	Degree of differentiation	The wavelength of the first extremum	The wavelength of the second extremum	The difference in the ratio
AK	6447/16	G1	288	334	3.2
MTC L/N	6436–37/16	G1	292	346	2.5
AK	9183/16	G2	290	335	1.7
AK	10132/16	G2	286	360	1.9
MTS omentum	10132/16	G2	287	342	2.0
MTS omentum	9273/16	G3	290	348	1.2
AK	1811/16	G3	292	340	1.4
AK	2765/16	G3	295	336	1.2
AK	207/19	G3	294	348	1.3
AK	25568/16	G4	296	340	0.8

The presented studies allow us to evaluate the possibility of using the spectroscopic fluorescence method in noninvasive surgery. The results of this work show that the use of fluorescence spectroscopy definitely has the potential to assess the condition of biological tissues, since there is a sensitivity of this method to the nature and degree of pathological changes.

Thus, it can be said that fluorescent diagnostic methods play an important role in biomedical research. With their help, the problem of noninvasive diagnosis of the pathological process in the tissue according to the degree of malignancy of the pathology can be solved.

References

1. Tahhan, N., Balanca, B., Fierstra, J., Waelchli, T., Picart, T., Dumot, C., Eker, O., Marinesco, S., Radovanpvic, I., Cotton, F., Berhouma, M.: Intraoperative cerebral blood flow monitoring in neurosurgery: A review of contemporary technologies and emerging perspectives. Neurochirurgie **68**(4):414–425 (2022)
2. Zhan, H., Sun, C., Xu, M., Luo, T., Wang, G., Xi, G., Liu, Z., Zhuo, S.: Analysis of intraoperative microscopy imaging techniques and their future applications. Front. Phys. **10**, 991279 (2022)
3. Wang, S., Yang, M., Li, R., Bai, J.: Current advances in noninvasive methods for the diagnosis of oral squamous cell carcinoma: a review. Eur. J. Med. Res. **28**(1), 53 (2023)
4. Melikechi, N. (ed.).: Optical Spectroscopy and Imaging for Cancer Diagnostics: Fundamentals, Progress, and Challenges (2023)
5. Wilson, B.C., Eu, D.: Optical spectroscopy and imaging in surgical management of cancer patients. Transl. Biophotonics **4**(3), e202100009 (2022)

6. Kaniyala, M.S., Kistenev, Y.V., Borisova, E., Ivanov, D., Zakharova, O., Boyko, A., Vrazhnov, D., Gopal, D., Chakrabarti, S., Prasada, S., Mazumder, N.: Types of spectroscopy and microscopy techniques for cancer diagnosis: a review. Lasers Med. Sci. **37**(8), 3067–3084 (2022)
7. Esposito, C., Borgogni, R., Autorino, G., Cerulo, M., Carulli, R., Esposito, G., Conte, F.D., Escolino, M.: Applications of indocyanine green-guided near-infrared fluorescence imaging in pediatric minimally invasive surgery urology: a narrative review. J. Laparoendoscopic Adv. Surg. Tech. **32**(12), 1280–1287 (2022)
8. Kalinina, S., Freymueller, C., Naskar, N., von Einem, B., Reess, K., Sroka, R., Rueck, A.: Bioenergetic alterations of metabolic redox coenzymes as NADH, FAD and FMN by means of fluorescence lifetime imaging techniques. Int. J. Mol. Sci. **22**(11), 5952 (2021)
9. Zherebtsov, E.A., Potapova, E.V., Mamoshin, A.V., Shupletsov, V.V., Kandurova, K.Y., Dremin, V.V., Abramov, A.Y., Dunaev, A.V.: Fluorescence lifetime needle optical biopsy discriminates hepatocellular carcinoma. Biomed. Opt. Express **13**(2), 633–646 (2022)
10. Deng, S., Gu, J., Jiang, Z., Cao, Y., Mao, F., Xue, Y., Wang, J., Dai, K., Liu, K., Wu, K., He, Q., Cai, K.: Application of nanotechnology in the early diagnosis and comprehensive treatment of gastrointestinal cancer. J. Nanobiotechnol. **20**(1), 415 (2022)
11. Belia, F., Biondi, A., Agnes, A., Santocchi, P., Laurino, A., Lorenzon, L., Pezzuto, R., Tirelli, F., Ferri, L., D'Ugo, D., Persiani, R.: The use of indocyanine green (ICG) and near-infrared (NIR) fluorescence-guided imaging in gastric cancer surgery: a narrative review. Front. Surg. **9**, 880773 (2022)
12. Knospe, L., Gockel, I., Jansen-Winkeln, B., Thieme, R., Niebisch, S., Moulla, Y., Stelzner, S., Lyros, O., Diana, M., Marescaux, J., Chalopin, C., Köhler, H., Pfahl, A., Maktabi, M., Park, J.H., Uang, H.K.: New intraoperative imaging tools and image-guided surgery in gastric cancer surgery. Diagnostics **12**(2), 507 (2022)
13. Marcus, C., Feizi, P., Hogg, J., Summerfield, H., Castellani, R., Sriwastava, S., Marano, G.D.: Imaging in differentiating cerebral toxoplasmosis and primary CNS lymphoma with special focus on FDG PET/CT. Am. J. Roentgenol. **216**(1), 157–164 (2021)

Structure of a Model of Situational Control for Intelligent Robots Containing SEMS Knowledge Bases, by Means of Coordinating Contexts of a Specialized Language of Their Communication

Alexander Ya. Fridman and Boris A. Kulik

Abstract *Problem statement*: In some previous collections of this series "Smart Electromechanical Systems", the authors proposed an information technology for distributing tasks in arbitrarily structured (in general case, network-centric) teams of robots based on analysis of their situational awareness. This paper considers the next stage in usage of such teams, namely the structural modifications necessary to organize interactions of the robots directly in the course of performing a common task. *Purpose of research*: For operative coordination of joint actions of robots, it is required to ensure their communication and mutual understanding in some common limited subset of natural language. The purpose of this stage of research is to develop and justify an appropriate method for providing information exchange between robots in the course of their work. *Results:* We propose to structure a robots' team as a collective of intelligent dynamic systems after G.S. Osipov and formalize the operative distribution of tasks among the robots as a constraint satisfaction problem that is solving at changes of the situation occurred in the course of the team work. A context-oriented method is proposed for describing the current situation in the language of communication of robots and for promptly adjusting their task parameters for each member of the team. The peculiarity of the method lies in the fact that the team coordinator forms contexts that must be mutually consistent for robots directly related to the progress of the task. *Practical significance*: The presented approach will allow to maintaining coordinated functioning of a team of robots in case of random (not caused by conscious opposition) deviations of the current situation from the nominal one, for which the tasks were originally distributed.

A. Ya. Fridman (✉)
Institute for Informatics and Mathematical Modeling, Kola Science Centre of RAS, 24A Fersman Str., 184209 Apatity, Russia
e-mail: fridman@iimm.ru

B. A. Kulik
Institute for Problems in Mechanical Engineering of the Russian Academy of Sciences, Bolshoy Pr. V.O., 61, 199178 Saint-Petersburg, Russia

I. L. Tarasova and B. A. Kulik (eds.), *Smart Electromechanical Systems*, Studies in Systems, Decision and Control 544, https://doi.org/10.1007/978-3-031-64277-7_11

Keywords SEMS-based robot · Robots cooperation · Coordination of robots' interactions · Normal and emergency operation mode

1 Introduction

In [1–3], we proposed an information technology for distributing the tasks within an arbitrarily structured (in general case, network-centric) team of robots on the basis of their situational awareness (SA) (see [4], for example). Below, the task of operative control for such teams is considered, which is necessary to regulate their interactions directly in the course of performing a common task. As goals of the team work and methods to formalize robots' descriptions can vary depending on many unpredictable factors, it looks rational to develop a flexible software environment that can be detailed and concretized for any reasonable specific case. With this in mind, we propose to operatively coordinate joint actions of robots with using some common limited subset of natural language, namely a set of propositions in the sufficiently general form IF–THEN-ELSE that is traditional for the production intelligence systems [5]. In our opinion, this is an appropriate method for organizing information exchange between robots and coordination of robots' actions in the course of their common work.

Our second proposal is to build software for teams of SEMS robots within the paradigm of intelligent dynamic systems (IDSs) (e.g., [6]) as it totally corresponds to the requirements of flexibility and generality we apply to the collectives of robots intended for doing something together. Besides, IDSs' software presupposes dynamic tasks, usage of knowledge bases to formalize intelligent behavior and, moreover, there exists a class of IDSs exploiting productions.

Thus, below we summarize our previous developments in managing teams of robots, brief basics of IDSs and eventually formulate our ideas in operative control of such teams. The general means to implement these proposals is provided by the situational control system (SCM) that is most completely described in [7].

2 SA for Teams of Robots

The SA concept soundly presupposes that any action becomes reasonable only after comprehending a current situation and making decisions on its necessary modification. The SCM provides a fairly strict formalization for the mathematical content of these three stages of the SA attaining with the opportunity to quantitatively estimate the gained degrees of the SA achieving [8] within a specific state space with an expertly determined metric [9] where the dangerous and critical states environ the admissible states of the system under modeling. With this goal, the non-dangerous (safe) range (SR) of the values for the state variables can be expanded, if necessary, with undesirable states, for which experts assign some relative degrees of danger.

Consequently, the emergencies and accidents can be modeled by the same apparatus that is used for modeling the normal states of the system and mathematically treated as some extensions of acceptable modes. For the multilevel (hierarchical) models, which are often used to reduce the complexity of the model, it makes also possible to indicate the moments of transitions from a normal to a dangerous state and reveal some root causes of such transitions by the detecting the lowest (in the hierarchy) object whose functioning breaks its SR. Besides, a decision-maker (DM) can analyze the degree of losing the SA in these transitions that is very useful for estimating the real situation in the progress of the entire task. This way we have provided usage of the "classical" conceptual spaces (e.g., [10–12]) for unacceptable modes as well.

The stability of performance is another important feature of the system quality. In the SCM, any sharp change in the values of the quality criteria the components of the system is treated as a sign of some sudden and therefore questionable behavior of the system. To analyze and estimate such changes, we propose to use increments (gradients) of these quality criteria as some additional means for determining the admissible time period, for which we can trust to the forecasts of the behavior for the corresponding component of the model: the larger the gradients grow, the smaller is this reliable period.

The implementation of these ideas for groups of robots requires for some modifications due to the specificity of such tasks [13, 14]. To transform these ideas for the SCM when all components of the system (robots) are the DMs, in [2] we analyzed both peer-ranked and hierarchical robots' teams. In the former case, it is possible to immediately apply the formulas for the attained degrees of the SA, which were proposed in [8], but here it is difficult to distribute tasks among the robots in order to organize their common activity in a reasonable way according to the principles of applying the swarm algorithms (see, for example, [15]). That is why we prefer to use hierarchies where principles of coordinating the elements' activities were developed long ago [16]. According to these Founding Fathers, a hierarchy of any complexity can be always treated as a two-level one where all "ordinary" DMs, each of which has its own (local) goal (the so-called "quality criterion"), are placed on the lower level, while the upper level contains the sole Coordinator, i.e., a higher-ranked DM that is responsible not only for reaching its personal goal, but also for coordinating the efforts of the lower DMs as to make them do their best for achieving the general (global) goal, for extremization of which the whole hierarchy was created. Who and why had formulated this global goal, remains totally secret. This approach is still relevant, so we developed some techniques to coordinate such structure based on the gradients of the quality criteria for both the Coordinator and the local DMs. Initially, all the latter ones were supposed to have a peer rank, but lately the techniques were generalized for arbitrary ranks of the lower-level DMs; each of them can have any subordinated structure [3]. Thus, the total configuration of the system (the entire group of robots in our case) can be arbitrary.

When the size of the group grows, any single hierarchical control structure becomes clumsy and poorly manageable, so we provided a possibility to apply the net-centric approach (see [17, 18], etc., etc.). We proposed that a net of any complexity can be formed, if every node of this net can have a hierarchical structure

[3], and developed some techniques to use the SA for estimating the behavior of such a system. In this interpretation, the major difference between hierarchical and net-centric structures is in forming areas of responsibilities (AoR) for the involved DMs, that is, in defining the lists of their subordinated robots. In hierarchies, this list is formed for every DM during consolidation of a hierarchy and stays fixed until the structure is changing, while for the net structures, any AoR can be operatively reformed if its DM is losing its degree of the SA. The AoR for every DM in the net includes all connected to it nodes whose degree of the SA is less than the DM has achieved currently. This idea specifies the miraculous process of the "migration of the centre" that is always declared but rarely justified in the net-centric applications. The only restriction for applying this technique is that, for any step of decision-making, the structure of the entire net must have no redundancy, that is, any possible structural alternatives have to be resolved prior to forming the AoRs. Hence, similar to the computational networks in the sense of Enn Tõugu [19], the entire system can be described as a graph with alternative subgraphs. This restriction does not look too burdening as the current general structure of the system (choosing the preferable structural alternatives) is decided by other algorithms during the classification of situations [7].

As soon as the mentioned restriction is met, the achieved degree of the SA is calculated for every node on the upper network level; the nodes with locally highest degrees are chosen as DMs and become responsible for coordination (scattering the tasks) of the nearest nodes with some lower degrees of the SA. In such networks, the distance can be calculating either by the number of the intermediate nodes (for information systems) or by the Euclidean metric (for material flows). For the robots, the second option fits better, though it is problem-dependent in general. Usually, the gradient coordination [7] can be useful as well. If a node can be included in several AoRs, some additional criteria or the random choice is used. As it was already mentioned, the AoRs are reformed in the case of any significant losses in the current degrees of the attained SA for any DMs.

However, the problem of a flexible transferring the higher-level requirements to the lower DMs was still unresolved because any Coordinator could change the tuning parameters of the quality criteria for the subordinated DMs, but had no mechanism to inform them any certain instructions. To do so and thus stipulate some intelligent interfaces among the DMs, below we propose to consider the DMs as the IDSs [6].

3 Robots as IDSs

An IDS is defined as an non-stationary system on the discrete time scale

$$D = \langle X, N, \varphi, \psi \rangle, \tag{1}$$

where: X is a topological space of possible IDS's states $x \in X$ and the distance relation σ; the set N includes natural numbers marking some points of the discrete time. Let ϑ denotes the set of all subsets for the set X;

$\varphi: \vartheta \to \vartheta$ means a closure function that allows to enrich the sets of the states for a current time step and has the property as follows:

$$x \subseteq \varphi(x) \text{ for any } x \in \vartheta; \tag{2}$$

a transition function $\psi: \vartheta \times N \to \vartheta$ has the following properties, which comply with the corresponding requirements of the traditional control theory:

$$\begin{aligned} &\psi(x, 0) = x \text{ for any } x \in \vartheta, \\ &\psi(\psi(x, t_1), t_2) = \psi(x, t_1 + t_2). \end{aligned} \tag{3}$$

Further on, we will consider the IDSs based on rules, which are composed of some formulas of a certain language L and has the state equation as:

$$x(t+1) = \varphi(\psi(\varphi(x(t) \cup u(t) \cup \delta(t))), \tag{4}$$

where: $u(t) \subseteq U(t) \subseteq U \subseteq L$ is a set of new facts (control signals) that appeared at a current step and need to be added to the former state $x(t)$; $\delta(t) \subseteq \Delta(t) \subseteq \Delta \subseteq L$ is a set of facts resulted from some disturbances—unexpected changes in the environment of the IDS.

Equation (4) is proved to have stable solutions for any monotonic φ and for any ψ that is monotonic on the state vector $x(t)$. Below, we consider all these conditions to be met.

3.1 Architecture of IDSs

In a rule-based IDS, the knowledge base (KB) includes a set of such rules:

$$\Pi = \langle C, A, D \rangle, \tag{5}$$

where: C is a condition to use a rule;
A lists the facts to be added after the running of the rule Π;
D includes the facts to be deleted after the running of the rule Π.
C, A and D are some logical formulas of L.

Any rule has to satisfy the correlation

$$A \cap D = \emptyset. \tag{6}$$

and belongs to a single class: either RD or RS.

Any RD-rule describes an action on the environment to be executed a variable is assigned with some admissible values of those database attributes, which are available for the current state. Such actions change the current state of the IDS's database. The RD-rules describe the temporal changes of the state $x(t)$ and are correspondingly called "transition (diachronic) rules".

RS-rules include no actions and do not impact the environment; rather, they describe the available knowledge of it, i.e., they formalize the subject domain's theory.

Hence, the KB of an IDS looks like

$$R=\langle \text{RS, RD} \rangle. \tag{7}$$

3.2 Goal-Driven Behaviour of the IDSs

Thus, an IDS is described by the formulas (1)–(4). To meet the requirement (2), RS-rules may not remove any facts (in them, the set D is empty).

If so, $\psi(\bullet, 1)$ is a function of transition, and an orbit (trajectory) of the IDS is defined by the set $\{\varphi(\psi(x, i))\}$ for $i \in N$.

In the Markov case, (4) describes the dynamics of a rule-based IDS.

Consider the connections between the architecture (7) and properties of the model (1)–(4).

Assume that $L(R)$ are formulas from L, which are used in the rules from R.

Definition 1. For a set X of some IDS's states, a pair of points (x_0, x_1) from the space $X \times X$ is N-attainable, if for all $j = 0, 1,..., N - 1$, such control signals $U(j)$ exist, for which, under the initial conditions $x(0)$ SYMBOL 205 \f "Symbol" \s 12H $x_0 \cup U(0)$, the relation x_1 SYMBOL 205 \f "Symbol" \s 12H $x(N)$ is met for some solutions $x(t)$, $t \in N$ of the Eq. (4).

Definition 2. For an N-attainable pair (x_0, x_1), if every fact from $x(N)$ is not repeated in several rules within the corresponding trajectory, such a pair (x_0, x_1) is effectively N-attainable.

Suppose, $\tau \subset L(R)$ is a set of facts. For a given sequence $\Pi_1, \Pi_2, ..., \Pi_k$ of R-rules, the set of the facts $S(\Pi_1, \Pi_2, ..., \Pi_k)$, which will be inferred after some τ rules taken from the sequence are applied, is defined by induction as follows:

$S(\Pi_1) = \tau \backslash D_1$ SYMBOL 200 \f "Symbol" \s 12И A_1;

$$S(\Pi_1, \Pi_2, \ldots, \Pi_i) = S(\Pi_1, \Pi_2, \ldots, \Pi_{i-1}) \backslash D(\Pi_i) \cup A(\Pi_i). \tag{8}$$

Definition 3. A rule Π_i is admissible, if there exists a control impact U_{i-1}, for which

$$C_i \subseteq (S(\Pi_1, \Pi_2, \ldots, \Pi_{i-1}) \cup U_{i-1}). \tag{9}$$

Definition 4. The sequence of rules and controls $\Pi = \langle(\Pi_1, U_1), (\Pi_2, U_2), \ldots, (\Pi_k, U_k)\rangle$ composes a plan for achieving the state ω starting from the current state ι, if:

(1) this sequence contains only admissible rules;
(2) ω SYMBOL 205 \f "Symbol" \s 12H $S(\Pi_1, \Pi_2, \ldots, \Pi_k)$.

Theorem 1. For any pair $(x_0, x_1) \in X \times X$, the plan $\Pi = \langle(\Pi_1, U_1), (\Pi_2, U_2), \ldots, (\Pi_k, U_k)\rangle$ exists, if and only if (x_0, x_1) is an N-attainable pair.

The definitions 1–4 and Theorem 1 that is proved in [6] constitute the brainware completely sufficient for transferring the Coordinator's preferences to the local (executing) robots at any step. The preconditions C contain the logical functions, which have to become true for applying a rule, while A and D describe the respective actions to be done, and the plan Π determines a sequence of rules to be applied for implementing the transition from x_0 to x_1.

The problem of building a language L suitable for accomplishing a certain task needs to be specified for every particular goal and composition of the robots spared to reach this goal. In general case, this language belongs to the category of the controlled natural languages (see, for example, [20]).

However, some general recommendations can be provided on how to avoid inconsistencies among the robots and consider contexts during the implementation of a plan. We propose to do it by using the paradigm of the constraint satisfaction problem (CSP) [21].

4 Contextual Constraint Management

The concept of "context" is used in linguistics, translation theory, system software (task/process context, query context), in the study of formal languages and grammars (context-free and context-sensitive languages), as well as in the development of decision support systems.

Within the framework of information technologies, the context is defined as information that specifies the description of the situation in which the object under study is currently located. In general, context allows determining what information is relevant to a particular situation (for solving a certain problem). Given the focus on intellectualizing technology, context includes not only information, but also knowledge that is relevant to the current task.

The context-oriented approach to constraint management in the SCM is based on the following classification of constraints: (1) the constraints that describe the designs allowed in the modeling system—modeling system constraints; (2) the restrictions, which feature the elements (types of elements) of the model that are used in the subject area under study – limitations of the subject area; (3) the restrictions inherent in the elements (types of elements) presented in a given fragment of the model (situation)—restrictions of the model fragment; (4) the restrictions on the values of variables formed during the simulation process within the framework of a preselected fragment of the model—restrictions at the simulation stage. For instructing the robots within a group, we can use constraints of all listed types except for the first one.

Let us list some advantages of the context-oriented constraint management in conceptual modeling systems.

1. The ability to activate only those contextual restrictions relevant for the domain model that is currently being studied. This allows to flexibly reconfigure and quickly analyze both the constraints common to the entire class of admissible models, and those typical to a specific subject area (model fragment, simulation step), which help to reduce the complexity of solving problems.
2. Based on the analysis of contexts at the model construction stage, more detailed control of the correctness of its structure, composition and correct connection of calculation modules is provided.
3. At the simulation stage, by comparing unplanned queries and contextual restrictions, incorrect calls to the database of the modeling system are monitored.

Application of the context-oriented approach to constraint processing made it possible to combine the advantages of the situational conceptual modeling and constraint programming, as well as to realize these advantages when solving modeling problems.

To formalize contexts, we apply the apparatus of n-tuple algebra (NTA) [21] briefly introduced below.

4.1 CSP Formalization in NTA

According to [22, 23], the CSP is defined by a set of *variables* $x_1, x_2, \ldots, x_n$, each of which has a non-empty *domain* D_i, and by a set of *constraints* $C_1, C_2, \ldots, C_m$. Each constraint C_i specifies a permissible combination of variables values in a certain projection, in some cases this can be expressed as a prohibition of certain combinations. A specific CSP solution is all (or at least some) combinations of variables values that satisfy all constraints.

Many well-known tasks can be reformulated as CSPs. These include, in particular, problems of solving systems of equations, optimization problems, graph coloring problems, propositional satisfiability problems (SAT), etc. In quite a few cases, such constraints can be expressed directly in the form of NTA-objects. The possibilities of using NTA in CSP solving were first investigated in (Zuenko 2014).

The CSP solution methods can be divided into three classes depending on the used solution algorithms [22]. Within the first class, problems are solved by *backtracking depth-first* search algorithms that build a solution by extending a partial assignment of variables values step by step, with using various heuristics and applying reasonable backtracking strategies. The second class includes problems that are solved by using *constraints propagation* algorithms; these algorithms exclude from the search space some combinations of values that do not satisfy constraints. The third type incorporates problems solved by implementing *structural algorithms* that use information about the structure of the primary or dual graph of constraints for a problem. Algorithms of this class decompose the original CSP into loosely coupled subproblems that can be solved with using some methods of the previous two classes. Thus, structural algorithms can also be used in combination with other types of algorithms.

Fairly often, when solving CSPs, constraints with a finite domain of definition are reduced to a set of binary relations by introducing a sufficient number of auxiliary variables.

A number of studies on satisfying constraints with finite domains use the concept of a *global constraint*. The classic example of a global constraint is the *Alldiff* constraint, according to which all participating variables must have different values. This constraint is often applied in practice and, therefore, it is proposed as a built-in constraint in major commercial and research systems for constraint programming [24, 25].

4.2 The CSP and Its Relationship with Logical Inference

Many constraints can be transformed into *n*-ary relations expressed with using NTA-objects, which can be divided into two types in terms of their content, namely: *forbidding constraints*, which contain elementary tuples that should not participate in the assignment of variables values in a solution, and *permitting constraints*, whose elementary tuples can be used to form solutions. Forming of forbidding or permitting relations (or NTA-objects) from concrete constraints expressed in a natural or formal language is one of the important intermediate tasks of solving CSP in NTA, its solution can help to achieve a significant reduction in complexity of solving CSPs. Thus, in [26] it was shown how the usage of NTA methods in solving the SAT problem leads to a significant reduction in complexity of the solving algorithms.

As for robot's task scheduling, it is mostly based on two types of binary relations: "start before / later than" and "run simultaneously / not simultaneously". In [27], we show how to transform these constraints into some NTA-objects, which can be easily represented as production rules.

Formalization of a CSP with finite domains within NTA allows a new approach to the formulation and solution of this problem. In most cases, the constraints C_1, C_2, …, C_m can be formed as relations (NTA-objects), which contain domains of possible variables values specified in the formulation of the problem and expressed by sets of elementary tuples. If there are given (or are simpler to formalize) forbidding constraints, they are transformed into permitting constraints by the complement operation. Since all permitting constraints act jointly and, in their meaning, are conjunctions of possible ranges of values of variables, the search for a solution to the CSP problem from the point of view of the NTA consists of two stages:

(1) transformation of the constraints $C_1, C_2, \ldots, C_m$ into NTA-objects;
(2) computation of the generalized intersection of the transformed constraints.

At all stages of solving the problem, the methods developed in the NTA are used to reduce dimensions of constraints and the complexity of computations.

5 Conclusion

It is shown that usage of numerical estimates of the situation awareness makes it possible to quickly construct areas of responsibility for the robots considered as decision-making nodes in a net-centric system and thus improve coordination of interactions among such robots in cases when coordinating robots lose their SA.

A context-oriented method is proposed for describing the current situation in the language of communication of robots and for promptly adjusting their task parameters for each member of the team. The peculiarity of the method lies in the fact that the team coordinator forms contexts, which must be mutually consistent for the robots directly related to the progress of the task.

The developed operative coordination method includes the following sequence of steps.

Step 1. Preliminary expression of the timetable for performing a task in the production format IF-THEN-ELSE (strategic planning of the task).
Step 2. Transformation of the timetable into NTA-objects followed by checks for consistency of the plan as a CSP (correctness test).
Step 3. Formation of the robots' KBs in the IDS format (7) followed by the check of attainability (existence of a plan (8), feasibility test).
Step 4. During the accomplishing of the task, the step-by-step estimations of the impact of the unpredicted disturbances and backtracking to the previous steps depending on the current degree of violating the plan.

The presented approach will allow to maintaining coordinated functioning of a team of robots in case of random (not caused by conscious opposition) deviations of the current situation from the nominal one, for which the tasks were originally distributed.

Acknowledgements This work was carried out within the framework of the state task for the IIMM KSC RAS (R&D "Development of Theoretical, Organisational and Technical Foundations of Information Support for Management of Viability in Regional Critical Infrastructures in the Arctic Zone of the Russian Federation" No. FMEZ-2022-0023).

The work was carried out within the framework of the state assignment of the Ministry of Science and Higher Education of the Russian Federation for the Institute of Problems in Mechanical Engineering of Russian Academy of Sciences.

References

1. Fridman, A.: SEMS-based control in locally organized hierarchical structures of robots collectives. In: Gorodetskiy, A.E., Kurbanov, V.G. (eds). Smart Electromechanical Systems: The Central Nervous System. Series: Studies in Systems, Decision and Control, vol. 95, pp. 31–50. Springer, Heidelberg (2017)
2. Fridman, A., Kulik, B.: Assessment of situational awareness in groups of interacting robots. In: Gorodetskiy, A.E., Tarasova, I.L. (eds.) Smart Electromechanical Systems: Situational Control, pp. 25–33. Springer Nature Switzerland AG (2020)
3. Fridman, A., Kulik, B.: Task scheduling within robots' collectives of arbitrary structures. In: Gorodetskiy, A.E., Tarasova, I.L. (eds.) Smart Electromechanical Systems: Behavioral Decision Making, pp. 95–100. Springer Nature Switzerland AG (2021)
4. Endsley, M.R.: Final reflections: situation awareness models and measures. J. Cogn. Eng. Decis. Making **9**(1):101–111 (2015)
5. The Role of Production Systems in AI Evolution: Access mode: https://www.simplilearn.com/production-system-in-ai-article. Retrieved 22 Jan 2024
6. Osipov, G.: Intelligent dynamic systems. J. Sci. Techn. Inf. Process. **37**(5):259–264 (2010)
7. Fridman, A.: Situational Modeling: Definitions, Awareness, Simulation. Nova Science Publishers, Inc, USA (2023)
8. Kulik, B.A., Fridman, A.Ya.: Quantitative assessment of situational awareness in a conceptual modeling system. In: System Analysis in Design and Management: Collection of Scientific Papers of the XXIII International. Scientific-Practical Conference Part 3. Publishing house Polytech-Press, St. Petersburg (2019) (In Russian)
9. Fridman, A.Ya., Kurbanov, V.G.: Situational modeling of reliability and safety of industrial-natural systems. J. Inf. Control Syst. **4**(71), 1–10 (2014) (In Russian)
10. Sowa, J.F.: Conceptual Structures—Information Processing in Mind and Machines. Addison-Wesley Publ. Comp. (1984)
11. Gärdenfors, P.: Conceptual Spaces: The Geometry of Thought. A Bradford Book. MIT Press, Cambridge, MA (2000)
12. Zenker, F., Gärdenfors, P.: Applications of Conceptual Spaces. The Case for Geometric Knowledge Representation. Synthese Library, vol. 359. Springer, Heidelberg (2015)
13. Endsley, M.R., Jones, W.M.: A model of inter- and intrateam situation awareness: Implications for design, training and measurement. In: McNeese, M., Salas, E., Endsley, M. (eds.) New trends in Cooperative Activities: Understanding System Dynamics in Complex Environments, pp. 46–67. Human Factors and Ergonomics Society, Santa Monica, CA (2001)
14. Ermolov, I.L.: Emerging issues of robots to be used in groups. In: Gorodetskiy, A.E., Kurbanov, V.G. (eds.) Smart Electromechanical Systems: Group Interaction. Series: Studies in Systems, Decision and Control, vol. 174. Springer Nature Switzerland AG, pp. 3–8 (2019)
15. Ab Wahab, M.N., Nefti-Meziani, S., Atyabi, A.: A comprehensive review of swarm optimization algorithms. PloS one **10**(5) (2015)
16. Mesarovic, M.D., Macko, D., Takahara, Y.: Theory of Hierarchical Multilevel Systems. Academic Press, New York–London (1970)
17. Cebrowski, A.K., Garstka, J.J.: Network-centric warfare: its origin and future. In: U.S. Naval Institute Proceedings. Annapolis (Maryland) (1998)
18. Kapustyan, S.G., Dyachenko, A.A.: Distributed information management system for automated multi-robot technical transport and warehouse complex. Mechatron. Autom. Manage. **7**, 34–39 (2012). (in Russian)
19. Tõugu, E.H.: Conceptual Programming. Nauka, Moscow (1984)
20. Kuhn, T.: A survey and classification of controlled natural languages. Comput. Linguist. **40**(1), 121–170 (2014)
21. Kulik, B., Fridman, A.: Complicated Methods of Logical Analysis Based on Simple Mathematics. Cambridge Scholars Publishing, Newcastle upon Tyne (2022)
22. Ruttkay, Z.: Constraint Satisfaction—A Survey. CWI Quarterly **11**, 163–214 (1998)

23. Russel, S., Norvig, P.: Artificial Intelligence: A Modern Approach, 2nd edn. Prentice Hall (2003)
24. Zuenko, A.A.: Constraint inference upon matrix representations of finite predicates. Artif. Intell. Decis.-Making **3**, 21–31 (2014). (in Russian)
25. Beldiceanu, N.: Global Constraints as Graph Properties on Structured Network of Elementary Constraints of the Same Type. Technical Report T2000/01, SICS (2000)
26. Kulik, B.A.: New classes of conjunctive normal forms with a polynomially recognizable property of satisfiability. Autom. Remote. Control. **56**(2), 245–255 (1995)
27. Fridman, A.Y., Kulik, B.A.: Analysis of relationships between problems of finding "Interesting" consequences and constraints satisfaction for SEMS knowledge bases in the framework of N-tuple algebra. In: Gorodetskiy, A.E., Tarasova, I.L. (eds.) Smart Electromechanical Systems. Studies in Systems, Decision and Control, vol. 419, pp. 13–26. Springer, Cham (2022). https://doi.org/10.1007/978-3-030-97004-8_2

Identification of the SEMS Module Using the Example of a Model of a Linear Electric Drive of a Counter-Reflector Actuator Taking into Account Constrains

Andrey Yu. Kuchmin

Abstract *Problem statement*: Radio telescopes are large, complex electromechanical systems designed to study radiation from cosmic radio sources. The study of extremely low intensities of radio signals from distant sources requires the creation of large fully rotating radio telescopes, which are subject to various types of influences, which significantly reduces the quality of observations. Modern radio telescopes are built using the principle of adaptive optics, when reflecting surfaces can change their shape and move in space to compensate for the influence of external influences. SEMS modules are widely used to solve problems. Correction of pointing errors of the dish system can be effectively performed by moving the secondary dish of the radio telescope, which is called a counter-reflector. Synthesizing a counter-reflector control system is a complex task, as it is necessary to take into account a huge number of various kinds of constraints. A counter-reflector with an actuator is a complex electromechanical system, which requires the development of a model taking into account the constraints and identification of its parameters. *Purpose*: Development of a model of a linear electric drive for the counter-reflector actuator of a millimeter-wave radio telescope, taking into account limitations. Creation of a method for identifying the developed model taking into account constraints. *Results*: Models of the linear electric drive of the counter-reflector of a millimeter-wave radio telescope have been developed and methods for identifying these models, taking into account constraints, have been proposed. *Practical significance*: The developed models and methods for their identification can be used to create a control system for the dish system of a millimeter-wave radio telescope.

Keywords SEMS · Control · Model · Dynamics · Identification · Constraint

A. Yu. Kuchmin (✉)
Institute for Problems in Mechanical Engineering of the Russian Academy of Sciences, Bolshoy Pr. V.O., 61, Saint-Petersburg 199178, Russia
e-mail: radiotelescope@yandex.ru

I. L. Tarasova and B. A. Kulik (eds.), *Smart Electromechanical Systems*, Studies in Systems, Decision and Control 544, https://doi.org/10.1007/978-3-031-64277-7_12

1 Introduction

The problem of identifying a SEMS module is considered using the example of a linear drive of a counter-reflector (CR) actuator using a modified identifier based on method of least squares (LS identifiers) with constraints for quasi-stationary systems, described in [1].

A counter-reflector is a secondary mirror in mirror systems of radiotelescopes and antennas [2–4]. The purpose of the CR is to redirect radio radiation collected by the main mirror (reflector) to the secondary focus area where the receiving equipment is installed. The use of adaptive CR makes it possible to compensate for the influence of deformations and pointing errors of the main mirror (MM) [5]. The structure of adaptive CRs provides for several levels of adaptation. The first of them is the installation of the CR on a special actuating mechanism [2–4], which provides movement along five degrees of freedom: three linear and two angular coordinates. The second level of adaptation is the implementation of the CR from separate panels, which allows you to change the surface profile of the CR, thus compensating for the deformations of the MM and the CR itself.

Electromechanical systems of parallel structure, the main element of which are linear electric drives (actuators) [6–10], have become widespread as actuating mechanisms of adaptive control systems. The use of such systems is associated with their compactness, the ability to provide movement in many degrees of freedom, as well as the ability to move large loads with high positioning accuracy.

Such systems consist of a movable platform to which the CR is attached, a system of linear electric drives and a base located on the crossbar of the radiotelescope. The crossbar is an element of the truss structure of a radiotelescope or antenna on which the actuator of the CR is mounted. Linear electric drives are connected to the platform and base using special hinges. The control panels of CR are also moved using linear electric drives.

When moving the adaptive CR, linear electric drives operate in different loading modes depending on the orientation of the radiotelescope and the CR. A change in the rigidity characteristics of electric drives under different types of loading and from the elongation of their rods is also observed, which significantly affects the positioning accuracy of the drive. Therefore, the task of identifying the parameters of these electric drives is actual.

2 Description of the Linear Electric Drive Model and Method for Identifying Its Parameters

The linear electric drive (LE) consists of two hinges designed for attachment to the base and platform, a DC electric motor, a planetary gearbox and a ball screw, a tension system that prevents the opening of the backlash.

An LE with the following characteristics is considered.

Electric Motor Parameters:
Armature winding time constant: 0.1 s/rad
Armature winding resistance: 0.625 Ω
Motor torque constant: 0.5 N*m/A
Back EMF coefficient: 0.06 V*s/rad
Motor armature moment of inertia: 0.0003 kg*m^2
Coefficient of viscous friction in bearings: 0.01 N*m*sec/rad
Supply voltage: ± 15 V.

Parameters of the planetary gearbox—ball screw system:
Damping coefficient: 1.4*10^5 N*sec/m
Gear ratio: 2*pi/0.0001 rad/m
Stiffness coefficient: 2*10^6 N/m
Weight of transported mass: 200–1200 kg

Based on the given data, a model of a linear electric drive was developed in Matlab/ Simulink, used to test the LS identifier with constraints for quasi-stationary systems. The model consists of three main modules:

- a block simulating a DC motor;
- a block simulating a gearbox and a ball screw;
- a block simulating a moving platform.

The model makes it possible to describe various loading options for a linear electric drive at different positions of the CR crossbar. When testing the identifier, the deviation of the linear electric drive parameters was simulated in the range of 50% from the nominal values.

The signal of the elongation rate of the LE rod was chosen as the signal used in the LS identifier. The control signal is the voltage supplied to the electric motor.

The LS identifier uses a system model in the form of a difference filter with a fixed frequency of samples:

$$v[k] + \sum_{i=1}^{N} a_i v[k-i] = -\sum_{i=1}^{N} b_i u[k-i], \tag{1}$$

where v—measured rate of elongation of the LE rod; k—discrete time number; N—difference filter order; a_i and b_i—difference filter coefficients; u—control signal. The vector of estimated coefficients is formed from the vectors of difference filter coefficients as follows $\mathbf{p} = [\boldsymbol{a}; \boldsymbol{b}]$, where $\boldsymbol{a}$ and $\boldsymbol{b}$—vectors composed of corresponding coefficients a_i and b_i.

To construct an LS identifier, it is necessary to estimate the dimension of the model used in it (1). This can be done using a simplified LE model.

Since model (1) is linear, consider a simplified LE model, including:

- linear model of a DC electric motor;
- linear model of the planetary gearbox—ball screw system;
- linear model of the moving platform.

This model makes it possible to estimate the order of the difference filter (1), the sampling period of time, and create a system of constraints for its coefficients, which significantly increases the stability of the identification process.

The differential equations of the simplified LE model in the state space have the form:

$$\dot{\mathbf{X}} = \mathbf{A_M X} + \mathbf{B_M} u,\ v = \mathbf{C_M X},\ \mathbf{X} = \left[I_{EM}\ \Omega_{EM}\ v\ d\ L \right]^{\mathrm{T}},$$

$$\mathbf{A_M} = \begin{bmatrix} -\frac{1}{T_{EM}} & -\frac{C_e}{T_{EM} R_{EM}} & 0 & 0 & 0 \\ \frac{C_m}{J_{EM}} & -\frac{D_{GB}}{J_{EM} I_{GB}^2} - \frac{\mu_{EM}}{J_{EM}} & \frac{D_{GB}}{J_{EM} I_{GB}} & \frac{C_{GB}}{J_{EM} I_{GB}} & 0 \\ 0 & \frac{D_{GB}}{m_{PL} I_{GB}} & -\frac{D_{GB}}{m_{PL}} & -\frac{C_{GB}}{m_{PL}} & 0 \\ 0 & -\frac{1}{I_{GB}} & 1 & 0 & 0 \\ 0 & 0 & 1 & 0 & 0 \end{bmatrix},\ \mathbf{B_M} = \begin{bmatrix} \frac{1}{T_{EM} R_{EM}} \\ 0 \\ 0 \\ 0 \\ 0 \end{bmatrix},$$

$$\mathbf{C_M} = \left[0\ 0\ 1000\ 0\ 0 \right], \tag{2}$$

where $\mathbf{X}$—state vector; $\mathbf{A}_{\mathrm{M}}$, $\mathbf{B}_{\mathrm{M}}$, $\mathbf{C}_{\mathrm{M}}$—matrices of model parameters (2); I_{EM}—motor current; Ω_{EM}—motor velocity; v—LE rod elongation rate; d—LE deformation; L—LE rod elongation; T_{EM}—armature winding time constant; R_{EM}—armature winding resistance; C_e—back EMF coefficient; C_m—motor torque constant; J_{EM}—moment of inertia of the electric motor rotor; D_{GB}—LE damping coefficient; C_{GB}—LE stiffness coefficient; I_{GB}—LE gear ratio; μ_{EM}—coefficient of viscous friction in bearings; m_{PL}—weight of the platform and transported mass.

The LE is identified by the rate of rod elongation, so the output integrator can be excluded from consideration. Thus, the order of the simplified model will be 4, which will correspond to the order of the difference filter of the LS identifier.

System (2) allows us to formulate a system of constraints imposed on the parameters of the difference filter (1). This can be done like this:

- it is necessary to move from a continuous system to a discrete one;
- obtain the transfer function of a discrete system from the input variable u to the output variable v;
- obtain dependences of changes in the coefficients of the transfer function of a discrete system on the ranges of changes in the parameters of the simplified model (2).

As constraints for the LS identifier, estimates of changes in the ranges of coefficients of the discrete transfer function obtained from system (2) can be selected. However, in this case, the constraints will be strictly tied to the structure of the model (2), which may affect the quality of identification of the real system. The best results are shown by constraints introduced as ratios of the coefficients of a discrete transfer function; in this case, by varying the maximum and minimum values of these ratios, you can obtain good identifier stability when working with real systems. To estimate the ranges of difference filter coefficient values (1) from the discrete transfer function obtained using (2), a program was developed in Matlab/Simulink, working according to the following algorithm:

Algorithm for estimating ranges of difference filter coefficients.

1. Data input: ranges of changes of system parameters (2);
2. Selecting the number of the coefficient of the discrete transfer function;
3. Formation of a system of interval constraints for the procedure for searching for the minimum and maximum of the selected coefficient;
4. Formation of a procedure for calculating the current value of the selected coefficient using system (2);
5. Calling the procedure for searching for the minimum and maximum of the selected coefficient;
6. If successful, then go to the next coefficient number (step 2), otherwise finish with an error message;
7. Generate ranges of variation of transfer function coefficients;
8. The end.

The ranges of variation of the transfer function coefficients make it possible to form a system of constraints for the coefficients **p** of the difference filter (1):

$$\lambda_{ji,\min} \le \frac{p_j}{p_i} \le \lambda_{ji,\max}, \; i = 1\ldots 2N-1, j = i+1\ldots 2N, \tag{3}$$

where $\lambda_{ji,\,\min}$, $\lambda_{ji,\,\max}$—the corresponding minimum and maximum values of the difference filter coefficient ratios. The system of constraints (3) must be supplemented with conditions on the signs of the coefficients **p**. The conducted studies showed that to identify LE it is enough to impose constraints at $i = 1$, $i = 2$ and on the signs of other coefficients. Then the generated system of restrictions will look like:

$$\begin{gathered} p_1 \le 0, \; \lambda_{j1,\max}p_1 - p_j \le 0, \; -\lambda_{j1,\min}p_1 + p_j \le 0, \; j = 2\ldots 2N, \\ -p_2 \le 0, \; -\lambda_{j1,\max}p_2 + p_j \le 0, \; \lambda_{j1,\min}p_2 - p_j \le 0, \; j = 3\ldots 2N, \\ p_3 \le 0, \; -p_4 \le 0, \; p_5 \le 0, \; p_6 \le 0, \; -p_7 \le 0, \; -p_8 \le 0. \end{gathered} \tag{4}$$

We reduce constraints (4) to the matrix inequality $\mathbf{Gp} \le 0$.

Let's form the goal function as described [1]. Let's move from Eq. (1) to the equation of sample vectors:

$$\mathbf{v} + \sum_{i=1}^{N} a_i \mathbf{A}^i \mathbf{v} = -\sum_{i=1}^{N} b_i \mathbf{A}^i \mathbf{u}, \tag{5}$$

where **v**—vector of samples of the measured signal of the LE rod extension ratio; **u**—sample vector of the measured control voltage signal on the electric motor; $\mathbf{A}^i$—time domain shift operator. Let's move on to the matrix representation of Eq. (5) and determine the model error $\boldsymbol{\varepsilon}$:

$$\begin{gathered}\mathbf{v} + \mathbf{F}\mathbf{p} = \varepsilon, \\ \mathbf{F} = \left[\mathbf{A}^1\mathbf{v},\ \mathbf{A}^2\mathbf{v},\ \mathbf{A}^3\mathbf{v},\ \mathbf{A}^4\mathbf{v},\ \mathbf{A}^1\mathbf{u},\ \mathbf{A}^2\mathbf{u},\ \mathbf{A}^3\mathbf{u},\ \mathbf{A}^4\mathbf{u}\right].\end{gathered} \tag{6}$$

On small data samples, LS identifiers are sensitive to the influence of the initial conditions of dynamic systems; to take this into account, it is necessary to exclude some of the equations from (6). This is done by multiplying by a special matrix **S**. Thus, Eq. (6) are transformed to the form:

$$\mathbf{S}\mathbf{v} + \mathbf{S}\mathbf{F}\mathbf{p} = \mathbf{S}\varepsilon. \tag{7}$$

Based on (7), we obtain an objective function, by minimizing which we determine the parameters of the difference filter of the LS identifier:

$$\begin{aligned}\Phi &= 0.5\boldsymbol{\varepsilon}^{\mathrm{T}}\mathbf{S}^{\mathrm{T}}\mathbf{S}\varepsilon = 0.5\left[\mathbf{S}\mathbf{v} + \mathbf{S}\mathbf{F}\mathbf{p}\right]^{\mathrm{T}}\left[\mathbf{S}\mathbf{v} + \mathbf{S}\mathbf{F}\mathbf{p}\right] = \\ &= 0.5\mathbf{p}^{\mathrm{T}}\mathbf{F}^{\mathrm{T}}\mathbf{S}^{\mathrm{T}}\mathbf{S}\mathbf{F}\mathbf{p} + \mathbf{v}^{\mathrm{T}}\mathbf{S}^{\mathrm{T}}\mathbf{S}\mathbf{F}\mathbf{p} + \mathbf{v}^{\mathrm{T}}\mathbf{S}^{\mathrm{T}}\mathbf{S}\mathbf{v} = \\ &= 0.5\mathbf{p}^{\mathrm{T}}\mathbf{H}\mathbf{p} + \Theta\mathbf{p} + c, \\ \mathbf{H} &= \mathbf{F}^{\mathrm{T}}\mathbf{S}^{\mathrm{T}}\mathbf{S}\mathbf{F},\ \boldsymbol{\Theta} = \mathbf{v}^{\mathrm{T}}\mathbf{S}^{\mathrm{T}}\mathbf{S}\mathbf{F},\ c = \mathbf{v}^{\mathrm{T}}\mathbf{S}^{\mathrm{T}}\mathbf{S}\mathbf{v}.\end{aligned} \tag{8}$$

Taking into account (4), written in matrix form, and (8), we formulate the synthesis of an LS identifier with constraints as a mathematical programming problem:

$$J = \min_{\mathbf{p}}\left\{0.5\mathbf{p}^{\mathrm{T}}\mathbf{H}\mathbf{p} + \Theta\mathbf{p} + c\right\},\ \text{with constraints } \mathbf{G}\mathbf{p} \le 0. \tag{9}$$

3 An Example of a Numerical Calculation of the Problem of Synthesizing an LS Identifier with Constraints.

Using an algorithm for estimating ranges of difference filter coefficient values for a simplified LE model, based on the structure of constraints (4), the matrix **G** was calculated. The constraint parameters (4) were chosen in such a way as to ensure a stable identification process over the entire range of changes in LE parameters.

$$
G = \begin{bmatrix}
1.0000 & 0 & 0 & 0 & 0 & 0 & 0 & 0 \\
6.5099 & -1.0000 & 0 & 0 & 0 & 0 & 0 & 0 \\
3.1484 & 0 & -1.0000 & 0 & 0 & 0 & 0 & 0 \\
0.2388 & 0 & 0 & -1.0000 & 0 & 0 & 0 & 0 \\
0.0287 & 0 & 0 & 0 & -1.0000 & 0 & 0 & 0 \\
0.0407 & 0 & 0 & 0 & 0 & -1.0000 & 0 & 0 \\
0.0420 & 0 & 0 & 0 & 0 & 0 & -1.0000 & 0 \\
0.0065 & 0 & 0 & 0 & 0 & 0 & 0 & -1.0000 \\
8.3699 & 1.0000 & 0 & 0 & 0 & 0 & 0 & 0 \\
2.4487 & 0 & 1.0000 & 0 & 0 & 0 & 0 & 0 \\
0.3070 & 0 & 0 & 1.0000 & 0 & 0 & 0 & 0 \\
0.0224 & 0 & 0 & 0 & 1.0000 & 0 & 0 & 0 \\
0.0316 & 0 & 0 & 0 & 0 & 1.0000 & 0 & 0 \\
0.0540 & 0 & 0 & 0 & 0 & 0 & 1.0000 & 0 \\
0.0084 & 0 & 0 & 0 & 0 & 0 & 0 & 1.0000 \\
0 & -1.0000 & 0 & 0 & 0 & 0 & 0 & 0 \\
0 & -2.6331 & 1.0000 & 0 & 0 & 0 & 0 & 0 \\
0 & -0.3301 & 0 & 1.0000 & 0 & 0 & 0 & 0 \\
0 & -0.0240 & 0 & 0 & 1.0000 & 0 & 0 & 0 \\
0 & -0.0340 & 0 & 0 & 0 & 1.0000 & 0 & 0 \\
0 & -0.0580 & 0 & 0 & 0 & 0 & 1.0000 & 0 \\
0 & -0.0090 & 0 & 0 & 0 & 0 & 0 & 1.0000 \\
0 & -3.3854 & -1.0000 & 0 & 0 & 0 & 0 & 0 \\
0 & -0.2567 & 0 & -1.0000 & 0 & 0 & 0 & 0 \\
0 & -0.0309 & 0 & 0 & -1.0000 & 0 & 0 & 0 \\
0 & -0.0437 & 0 & 0 & 0 & -1.0000 & 0 & 0 \\
0 & -0.0451 & 0 & 0 & 0 & 0 & -1.0000 & 0 \\
0 & -0.0070 & 0 & 0 & 0 & 0 & 0 & -1.0000 \\
0 & 0 & 1.0000 & 0 & 0 & 0 & 0 & 0 \\
0 & 0 & 0 & -1.0000 & 0 & 0 & 0 & 0 \\
0 & 0 & 0 & 0 & 1.0000 & 0 & 0 & 0 \\
0 & 0 & 0 & 0 & 0 & 1.0000 & 0 & 0 \\
0 & 0 & 0 & 0 & 0 & 0 & -1.0000 & 0 \\
0 & 0 & 0 & 0 & 0 & 0 & 0 & -1.0000
\end{bmatrix}.
$$

Next, this identifier was verified using a simplified LE model. For different standard impacts, matrices of goal function parameters were formed (8).

Figures 1, 2, 3, and 4 show examples of such impacts and the corresponding system response.

Calculation example 1

Matrices of parameters of goal function will be:

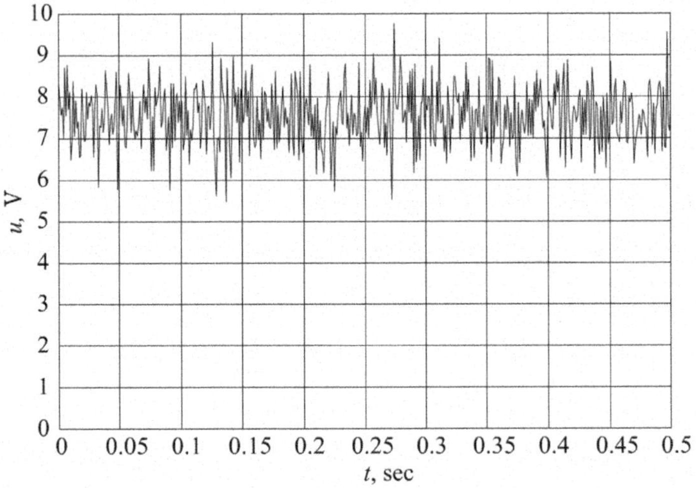

Fig.1 Input DC voltage + 7.5 V with added noise implementation with normal distribution and dispersion 0.5 V

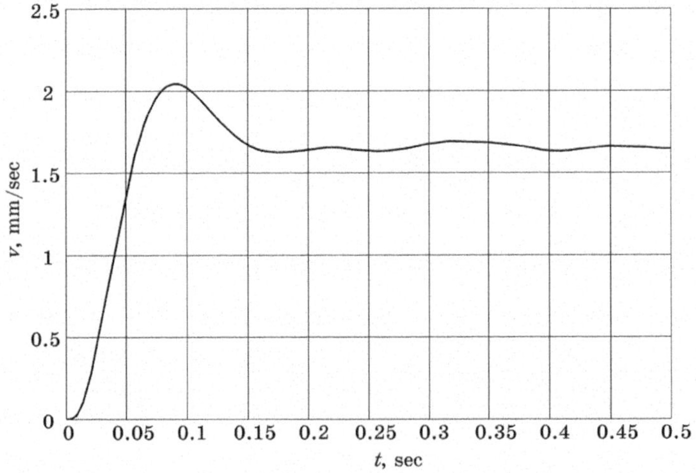

Fig.2 Measured rate of elongation of the LE rod mm/s for the input voltage shown in Fig. 1

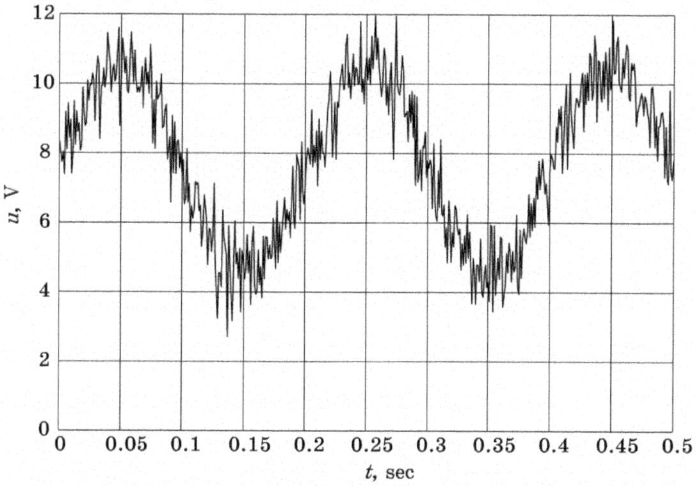

Fig. 3 Input shifted to + 7.5 V sinusoidal voltage with an amplitude of 3 V and a frequency of 5 Hz with the addition of noise implementation with normal distribution and dispersion 0.5 V

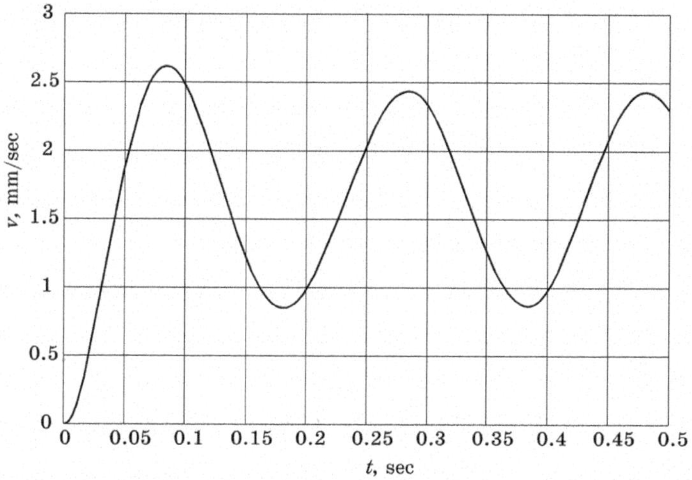

Fig. 4 Measured rate of elongation of the LE rod mm/s for the input voltage shown in Fig. 3

$\mathbf{H} = 1.0e + 004*$

[0.1326 0.1325 0.1324 0.1322 0.5946 0.5947 0.5947 0.5943
0.1325 0.1324 0.1322 0.1321 0.5934 0.5934 0.5935 0.5931
0.1324 0.1322 0.1321 0.1320 0.5921 0.5922 0.5922 0.5919
0.1322 0.1321 0.1320 0.1318 0.5909 0.5910 0.5910 0.5906
0.5946 0.5934 0.5921 0.5909 2.8457 2.8195 2.8200 2.8191
0.5947 0.5934 0.5922 0.5910 2.8195 2.8463 2.8202 2.8195
0.5947 0.5935 0.5922 0.5910 2.8200 2.8202 2.8473 2.8199
0.5943 0.5931 0.5919 0.5906 2.8191 2.8195 2.8199 2.8451],

$\boldsymbol{\Theta} = 1.0e + 003*$

$\left[1.3278\ 1.3264\ 1.3248\ 1.3233\ 5.9583\ 5.9589\ 5.9592\ 5.9558\right].$

It is necessary to find out whether the matrix **H** is positive definite; for this, all its angular minors must be positive. Calculations show that this condition is not met, so we are not dealing with a quadratic programming problem. To solve it, we used the complex method of Box [11], which is a further development of the Nelder-Mead simplex method, which allows solving optimization problems in the presence of constraints. The method of sequential quadratic programming was also used, where the matrix **H** is reduced to positive definite during the execution of the algorithm [12].

Comparison of parameter estimates obtained using the $\mathbf{p}_I$ identifier and the parameters of the simplified LE model $\mathbf{p}_M$ (Table 1):

Calculation example 2

Matrices of parameters of goal function will be:

Table 1 Identification results for calculation example 1

$\mathbf{p}_M$	$\mathbf{p}_I$
– 3.8439	– 3.8442
5.5401	5.5409
– 3.5483	– 3.5490
0.8520	0.8523
– 0.0796*1.0e–004	– 0.0795*1.0e–004
– 0.2285*1.0e–004	– 0.2286*1.0e–004
0.2286*1.0e–004	0.2289*1.0e–004
0.0729*1.0e–004	0.0729*1.0e–004

Table 2 Identification results for calculation example 2

$\mathbf{p}_M$	$\mathbf{p}_I$
− 3.4450	− 3.4432
4.3668	4.3625
− 2.3972	− 2.3942
0.4755	0.4749
− 0.0416*1.0e−003	− 0.0405*1.0e−003
− 0.0989*1.0e−003	− 0.1028*1.0e−003
0.1090*1.0e−003	0.1132*1.0e−003
0.0285*1.0e−003	0.0264*1.0e−003

$$\mathbf{H} = 1.0\mathrm{e} + 004*$$

$$\begin{bmatrix} 0.1575 & 0.1572 & 0.1569 & 0.1565 & 0.6778 & 0.6797 & 0.6816 & 0.6828 \\ 0.1572 & 0.1569 & 0.1566 & 0.1563 & 0.6741 & 0.6761 & 0.6780 & 0.6793 \\ 0.1569 & 0.1566 & 0.1564 & 0.1561 & 0.6705 & 0.6725 & 0.6744 & 0.6757 \\ 0.1565 & 0.1563 & 0.1561 & 0.1558 & 0.6667 & 0.6688 & 0.6707 & 0.6721 \\ 0.6778 & 0.6741 & 0.6705 & 0.6667 & 3.3576 & 3.3313 & 3.3315 & 3.3299 \\ 0.6797 & 0.6761 & 0.6725 & 0.6688 & 3.3313 & 3.3584 & 3.3321 & 3.3308 \\ 0.6816 & 0.6780 & 0.6744 & 0.6707 & 3.3315 & 3.3321 & 3.3592 & 3.3315 \\ 0.6828 & 0.6793 & 0.6757 & 0.6721 & 3.3299 & 3.3308 & 3.3315 & 3.3565 \end{bmatrix},$$

$$\boldsymbol{\Theta} = 1.0\mathrm{e} + 003*$$

$$\begin{bmatrix} 1.5771 & 1.5741 & 1.5708 & 1.5672 & 6.8139 & 6.8329 & 6.8510 & 6.8634 \end{bmatrix}.$$

In this case, the matrix $\mathbf{H}$ is also not positive definite.

Comparison of parameter obtained using the identifier $\mathbf{p}_I$ and the parameters of the simplified LE model $\mathbf{p}_M$ is shown in Table 2.

The LS identifier with constraints was compared with the classic LS identifier described in [13]. Using the classic identifier, it was not possible to achieve a stable estimate of the parameters when the dimension of the identifier model is equal to 4 when estimating the transfer function parameters of the simplified LE model. A stable and predictable process for estimating parameters was only possible when the dimension of the identifier model was equal to 2. Analysis of the frequency characteristics of the classical identifier model showed that this identifier only determines the parameters of the system of moving mass + elastic element of the LE and does not describe the motor. This is due to the poor computational conditionality of the matrices used in calculating the estimates of the parameters of the LE model. This is an expected effect due to the fact that due to the large gear ratio of the LE gearbox, there is a weak reaction to the electric motor when moving the platform with a load. For the LS identifier with constraints ensures stable identification when the model dimension is equal to 4; this weak link is taken into account in the reaction of the system of moving mass + elastic element of the LE to the electric motor.

4 Example of Identification of a Nonlinear Model of LE

The problem of identifying a nonlinear model of LE developed in the Matlab/Simulink environment was solved. The model consists of two main blocks:

- permanent electric motor block shown in Fig. 5;
- a block simulating the gearbox—ball screw—movable platform system.

There are the following symbols in Fig. 5: Ud denotes the control voltage supplied to the engine; Id—electric motor current; Wd—electric motor velocity; Ce—back-EMF coefficient; Rja—armature winding resistance; Tja—armature winding time constant; Cm—motor torque constant; Jmac—moment of inertia of the motor armature; Mumac—coefficient of viscous friction in bearings; s—designation of the function of a complex variable in the transfer functions of the corresponding blocks.

There are the following designations in Fig. 6: Ired—gear ratio; Dstr—damping coefficient of the gearbox—ball screw—moving platform system; Cstr is the stiffness coefficient of the gearbox—ball screw—moving platform system; Mstr—nominal equivalent mass of the ball screw pair—moving platform system; s—designation of the function of a complex variable in the transfer functions of the corresponding blocks.

To solve the problem of identifying the described nonlinear model of LE, an LS identifier with constraints is used. Finding the parameters of the identifier model is performed as a solution to the mathematical programming problem (9). The type of

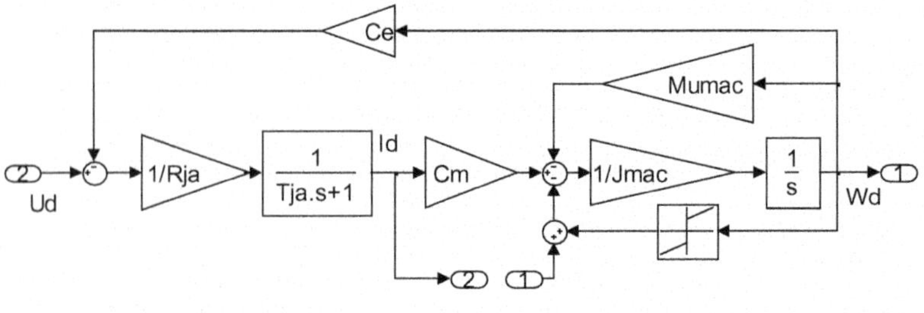

Fig. 5 DC motor model

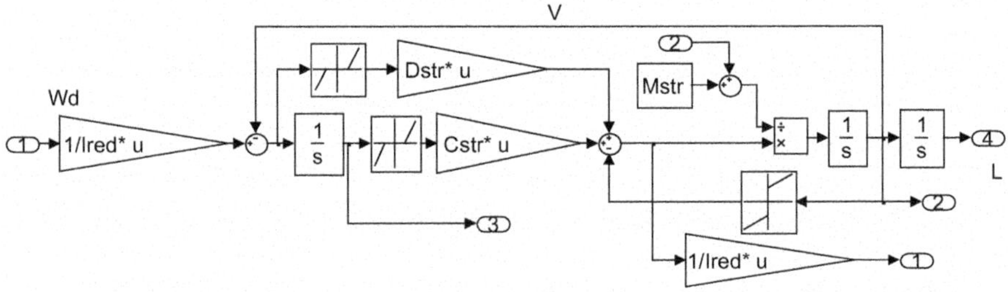

Fig. 6 Model of the gearbox—ball screw—moving platform system

setting voltage for the electric motor is shown in Fig. 3. The parameter matrices for (9) were obtained:

$$\mathbf{H} = 1.0\mathrm{e}+004*$$

$$\begin{bmatrix} 0.0991 & 0.0990 & 0.0988 & 0.0986 & 0.5284 & 0.5296 & 0.5309 & 0.5317 \\ 0.0990 & 0.0988 & 0.0986 & 0.0985 & 0.5258 & 0.5271 & 0.5283 & 0.5291 \\ 0.0988 & 0.0986 & 0.0985 & 0.0983 & 0.5232 & 0.5245 & 0.5257 & 0.5265 \\ 0.0986 & 0.0985 & 0.0983 & 0.0982 & 0.5206 & 0.5219 & 0.5231 & 0.5240 \\ 0.5284 & 0.5258 & 0.5232 & 0.5206 & 3.3576 & 3.3313 & 3.3315 & 3.3299 \\ 0.5296 & 0.5271 & 0.5245 & 0.5219 & 3.3313 & 3.3584 & 3.3321 & 3.3308 \\ 0.5309 & 0.5283 & 0.5257 & 0.5231 & 3.3315 & 3.3321 & 3.3592 & 3.3315 \\ 0.5317 & 0.5291 & 0.5265 & 0.5240 & 3.3299 & 3.3308 & 3.3315 & 3.3565 \end{bmatrix},$$

$$\mathbf{\Theta} = 1.0\mathrm{e}+003*$$

$$\begin{bmatrix} 0.9929 & 0.9911 & 0.9893 & 0.9874 & 5.3095 & 5.3223 & 5.3345 & 5.3425 \end{bmatrix}.$$

Matrix **H**, as in previous examples, is not positive definite. The parameters of the difference filter of the LS identifier have the values:

$$\mathbf{p}_I = \left[-3.0348, 3.3578, -1.5958, 0.2732, -0.0696 * 10^{-4}, -0.5749 * 10^{-4}, 0, 0\right].$$

Using the coefficients $\mathbf{p}_I$, we will construct an example of the transfer function of a nonlinear model of LE based on the rate of rod elongation:

$$W_{LE}(z) = \frac{V_{LE}(z)}{U_{EM}(z)} = \frac{0.0696z^{-1} + 0.5749z^{-2}}{1 + 3.0348z^{-1} - 3.3578z^{-2} + 1.5958z^{-3} - 0.2732z^{-4}} 10^{-4}, \quad (10)$$

where V_{LE} is the z-transformation of the velocity of movement of the LE rod; U_{EM}— z- transformation of the control voltage of the electric motor; z is a complex variable in the z-transform.

In Fig. 7 shows the results of identifying a nonlinear model of LE using an LS identifier with constraints. The dotted line shows the LE rod extension signal obtained from the nonlinear model of LE. The solid line indicates the LE rod extension signal obtained using the transfer function (10).

The studies show that, in comparison with the classical LS identifier, the proposed LS identifier with constraints provides greater stability when identifying dynamic objects, allows us to take into account weak connections in reactions in the structure of objects, which is difficult to organize in the classical LS identifier due to the poor conditionality of the matrices arising in this case. It is shown that the proposed identifier maintains the stability of the identification process in the presence of essentially nonlinear model elements, such as backlash and dry friction.

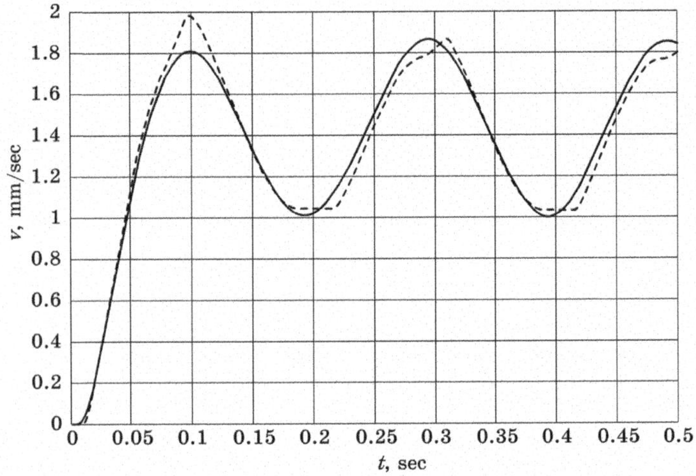

Fig. 7 Results of identification of the nonlinear model of LE. Measured rate of elongation of the LE rod mm/sec

5 Conclusions

The article analyzes the main problems that arise when identifying modern information, measuring and control systems of dynamic objects built on cybernetic principles. It is shown that in modern cyber-physical systems the identification process is an integral part of control loops. The influence of constraints on the process of identification that arise during the operation of such systems, which must be taken into account when synthesizing control laws, is described. As a result of the identification process, transfer functions of the linear electric drive were constructed from the input to the control voltage to the output signal—the rate of rod elongation. Using the example of identifying a linear electric drive, the advantage of LS identifiers with restrictions is shown when constructing models of systems with weak links in reactions between subsystems.

Acknowledgements The work was carried out within the framework of the state assignment of the Ministry of Science and Higher Education of the Russian Federation for the Institute of Problems in Mechanical Engineering of Russian Academy of Sciences.

References

1. Yu, K.A.: Creation of databases and formalization of knowledge using identification of SEMS, working under various constrains. In: Smart Electromechanical Systems: Mathematical and Software. Springer, Part 2, in this volume.

2. Artemenko, Y.N., Agapov, V.A., Dubarenko, V.V., Kuchmin, A.Y.: Co-operative control of Subdish actuators of radio telescope. Informatsionno-upravliaiushchie sistemy **4**, 2–9 (In Russian) (2012)
3. Gorodetskiy, A.E., Kurbanov, V.G., Tarasova, I.L., Kuchmin, A.Y.: Electric drives of system of logical control of the position of a counterreflector of the space radio telescope//Antenna **4**, 52–55 (In Russian) (2011)
4. Gorodetskiy, A.E., Kurbanov, V.G., Tarasova, I.L., Kuchmin, A.Y.: Structure of the system of logical control of the position of a counterreflector of the space radio telescope//Antenna **4**, 56–59 (In Russian) (2011)
5. Artemenko, Y.N., Gorodetsky, A.E., Dubarenko, V.V., Kuchmin, A.Y., Tarasova, I. L.: Problems of development of space radio-telescope adaptation systems. Informatsionno-upravliaiushchie sistemy **3**, 2–8 (In Russian) (2010)
6. Gorodetskiy, A.E., Kurbanov, V.G. (eds.): Smart Electromechanical Systems: The Central Nervous System, 266 p. Springer International Publishing AG (2017)
7. Gorodetskiy, A.E. (ed.): Smart Electromechanical Systems, 277 p. Springer International Publishing Switzerland (2016)
8. Merlet, J.P.: Parallel Robots (Solid Mechanics and Its Applications). Springer, Berlin (2004)
9. Heylo, S.V., Glazunov, V.A., Palochkin, S.V.: Handling mechanisms of parallel structure. Dyn. Anal. Manag., 86 (In Russian) (2014)
10. Artemenko, Y.N., Badu, E.I., Gorodetsky, A.E., Dubarenko, V.V., Kuchmin, A.Y.: Tarasova of I.L. Antenn of the radio telescope. Patent for the invention of RUS 2421765 (In Russian) (2010)
11. Bunday, B.D.: Basic Optimisation Methods, 128 p. E. Arnold, Baltimore, MD (1984)
12. Gill, P., Wong, E.: Convexification schemes for SQP methods//UCSD center for computational mathematics technical paper. CCoM-14-06, pp. 1–12 (2014)
13. Isermann, R., Münchhof, M.: Identification of Dynamic Systems. An Introduction with Applications, 705 p. Springer (2011)

Decision-Making

Adaptive Algorithms for Evaluating the Effect of Microdynamics on Crystal Growth Parameters in the Information-Control Technological Module of SEMS in Microgravity

Sergey N. Sayapin

Abstract *Problem statement*: despite half a century of experience in growing semiconductor crystals in space, no one in the world has yet managed to obtain crystals superior to the best terrestrial analogues. The reason for this is the statistical uncertainty and limited observability of crystal growth processes carried out on board spacecraft. In this regard, the question arose about the creation of an on-board information-control technology module (OBICTM) SEMS for growing high-quality crystals in microgravity conditions, capable of also performing the function of an on-board expert. This approach will improve the quality of the materials obtained by carrying out thermophysical processes, such as crystal cultivation, taking into account the recommendations from the on-board expert, issued in real time. The absence of such systems inevitably leads to the accumulation of raw scientific information from the director of the experiment about the course of the process on board the spacecraft. However, their creation requires solving a number of important tasks, one of which is the ability of OBICTM SEMS to carry out an operational assessment of the impact of vibration on the parameters of crystal growth in real time. *Purpose of research*: to solve the problem of operational measurement of the frequencies of microdynamic influences that affect the crystal growing process on board the spacecraft during the formation of control commands in the OBICTM SEMS, it is necessary to develop a new method and appropriate algorithms. *Results*: to achieve this goal, an adaptive algorithm for evaluating the effect of vibration on crystal growth parameters in real time in relation to OBICTM SEMS has been developed, and numerical simulation results have been obtained. To verify the developed adaptive algorithm and simulate OBICTM SEMS, a stand was made that simulates the conditions of cosmic microdynamics during crystal growth. *Practical significance*: the developed adaptive algorithms for evaluating microdynamics can be used as part of OBICTM SEMS for grow high-quality crystals in microgravity and provide the opportunity to obtain the necessary information on the basic physical parameters of the thermophysical process. The processed information will make it possible to deepen fundamental

S. N. Sayapin (✉)
Mechanical Engineering Research Institute of the Russian Academy of Sciences, Moscow, Russia
e-mail: S.Sayapin@rambler.ru

I. L. Tarasova and B. A. Kulik (eds.), *Smart Electromechanical Systems*, Studies in Systems, Decision and Control 544, https://doi.org/10.1007/978-3-031-64277-7_13

knowledge in the field of space materials science and improve existing technological processes on this basis.

Keywords Space materials science · Information technological systems · SEMS · Adaptive Algorithms · Method for evaluating the effect of microdynamics on crystal growth

1 Introduction

In recent years, work in the field of space materials science has intensified, in particular on the cultivation of solid crystals by astronauts on board the international space station [1, 2]. To reduce the level of micro-accelerations to (10^{-5}–10^{-6}) g in the crystal growing process in Russia, autonomous "Photon-M" spacecraft were used. Information about conducting experiments on "Photon-M" was transmitted via a telemetry channel to receiving stations [3, 4]. Thus, the methodology for studying the effects of microdynamics on the parameters of technological processes of crystal growth in microgravity was mainly reduced to recording micro-accelerations on board a spacecraft during space technological experiments, delivering recordings and the resulting sample of material to the experimental director and comparing the results.

Despite half a century of experience in growing semiconductor crystals in space, no one in the world has yet managed to obtain crystals that, in terms of their combined properties, would surpass the best terrestrial analogues [5]. In this regard, the question arose about the creation of space information technology systems for monitoring and controlling thermophysical processes (SITS MC TP) with elements of artificial intelligence, which also allow performing the function of an on-board expert. This approach will improve the quality of the materials obtained by conducting thermophysical processes, such as crystal cultivation, taking into account the recommendations from the on-board expert, issued in real time [6, 7]. At the same time, recommendations are formed on the basis of an operational assessment of the effect of microdynamic influences on crystal growth, on the one hand, and the possibility of reducing their impact on the process using an active spatial vibration protection system, on the other. In general, the block diagram of the SITS MC TP is similar to the block diagram of the universal module (UM) SEMS presented in the work (Fig. 1), which includes an electromechanical system (EMS) based on a parallel structure mechanism, an automatic control system (ACS), a measuring system (MS) and a docking system (DS) [8].

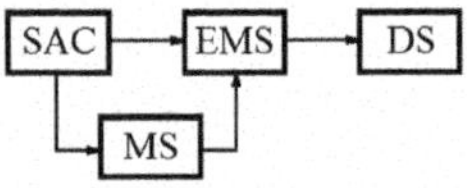

Fig. 1 Block diagram of the UM SEMS

Unlike the UM SEMS in SITS MC TP, the EMS unit is made in the form of an active low-frequency vibration-proof platform (if necessary with passive high-frequency vibration isolation) based on a parallel structure mechanism combined into one unit with technological equipment, for example, a crystal growing plant installed directly on the platform. The structure of such a system is similar to the structure of an intelligent active vibration protection and guidance system for precision large-sized space structures in real time, in which the precision structure is combined with an active vibration protection platform and also represents a single SEMS unit (Fig. 2) [9]. In general, the SITS MC TP includes an active vibration-proof platform in the form of the platform 1 itself and the base 2 mounted on the supporting structure 3 of the spacecraft. Platform 1 and base 2 are rigidly attached to the docking nodes by unit 4 of on-board technological equipment, for example, for growing high-quality semiconductor crystals.

The platform 1 is pivotally connected to the base 2 by means of a suspension with six degrees of freedom, made in the form of a manipulator, including at least six single active vibration isolation modules 5 of the same design and built on the basis of parallel connected drive kinematic circuits with the possibility of forming a spatial truss when the executive organs of the single active vibration isolation modules 5 are disconnected, each of which is equipped with a drive linear displacement with relative displacement sensor 6, allowing platform 1 to provide six-degree mobility relative to the base 2 and spatial positioning of platform 1 relative to the base 2. In the places of hinge attachment of each of the single active vibration isolation modules 5 to the platform 1 and base 2, respectively, combined sensors of spatial position and accelerations 7 and 8 are installed, made, for example, in the form of miniature three-axis

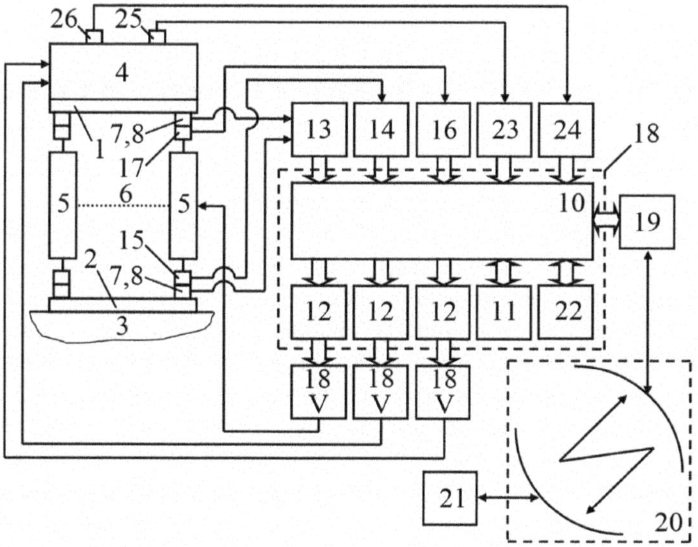

Fig. 2 Block diagram of the SITS MC TP

blocks of gyroscopes-accelerometers, allowing to determine the relative movements of single active vibration isolation modules 5 and measure the components of accelerations acting on their longitudinal axes. On-board high-performance information and measurement control system (OBHPIMCS) 9 includes a neurocomputer 10, software and algorithmic support 11 and digital-to-analog converters (DAC) 12. The inputs of the OBHPIMCS 9 are connected via the data buses of analog-to-digital converters (ADC), respectively, to the outputs of the ADC 13 of the combined spatial position and acceleration sensors 6 and 7; ADC 14 relative velocity sensors 15, ADC 16 relative displacement sensors 17, and the outputs of OBHPIMCS 9 are connected via output data buses to the corresponding inputs of series-connected DACs 12, power amplifiers 18 and actuators (single active vibration isolation modules 5, displacement drive and heater unit 4 of on-board technological equipment) and to the input of a series-connected on-board the radio telemetry system (BRTS) 19, the radio line 20 and the ground receiving and recording system 21. To improve the quality of the technological processes carried out, for example, for growing semiconductor crystals, BVIIUS 9 includes a unit for operational diagnostics of crystal growth 22, providing optimal modes of the technological process of crystal growth based on information received through ADC 23 and 24 from temperature sensors 25 and an ultrasonic sensor 26 for monitoring the position and shape of the crystallization front. In order to implement the effective functioning of the unit for operational diagnostics of crystal growth 22 and ensure optimal modes of the technological process of crystal growth, it is necessary to reduce statistical uncertainty and limited observability, leading to instability of the technological process and, as a result, unstable quality of the space technological experiment.

2 A Method for Assessing the Effect of Vibration on the Parameters of Technological Processes of Crystal Growth

To verify the developed adaptive algorithm, experiments were carried out on a stand (Fig. 3) modeling the conditions of cosmic microdynamics. Crystal growth in experiments was carried out with simultaneous registration of the spectrum of its main parameters (deflection and velocity of movement of the crystallization front) and the spectrum of microdynamic impacts. The modeling stand was organized on the basis of "SEISMOGEN-V-G" installations, the main purpose of which is the physical modeling of low-frequency vibrations of the earth's surface caused by earthquakes, explosions and other phenomena. The "SEISMOGEN-V-G" installations allow reproducing and measuring vertical and horizontal movements of the vibration platform table relative to the concrete foundation on which they are installed, as well as attenuating microseismic vibrations of the soil. The main technical characteristics of the installations are as follows:

- amplitude range of reproducible vibrations from 0.001 to 10 mm;

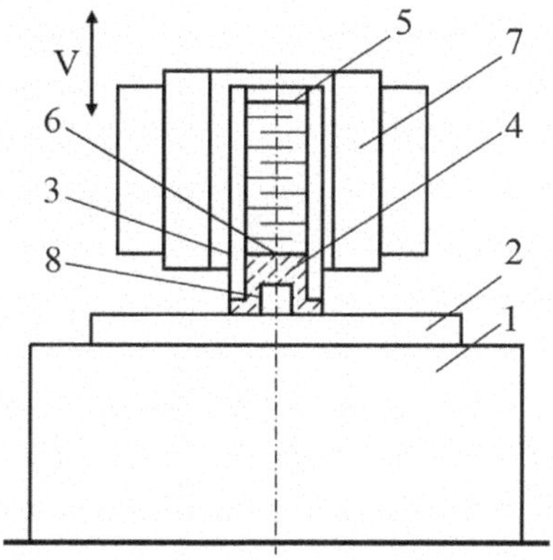

Fig. 3 Modeling stand of microdynamic impacts

- frequency range of reproducible vibrations from 0 to 20 Hz;
- the measurement error of the oscillation amplitude of the platform is from 1 to 3%;
- overall dimensions of the vibrating platform are 1000 × 800 mm;
- load capacity up to 50 kg.

The "SEISMOGEN-V-G" installations include a computer control and information processing system that ensures their operation, measurement of parameters of reproducible vibrations and generation and processing of vibrations of arbitrary shape.

The modeling stand (Fig. 3) is an installation 1 "SEISMOGEN", generating and controlling the specified microdynamic impacts, on a vibrating platform 2 of which a container 3 with a sample forming solid 4, liquid 5 parts and an interphase boundary (crystallization front) 6 is rigidly installed in a vertical position on a vibrating platform 2. The container is located inside an electric muffle furnace 7, capable of moving vertically relative to the sample container at set speeds V. To control the deflection (curvature) and the speed of movement of the crystallization front, an ultrasonic sensor 8 is installed on the outer end surface of the solid part of the sample, generating high-frequency pulses and receiving their reflection from the surface of the crystallization front 6. Increasing the accuracy of measuring the parameters of the technological process is achieved through appropriate statistical data processing using developed programs. The developed method of ultrasonic control of technological process parameters is similar to the methods described in [10, 11]. The process of directional crystallization is carried out by melting the upper part of the sample in the furnace, holding it and then moving it vertically at a predetermined speed (1–15 mm/h).

The complexity of solving the problem of assessing the mutual influence of the spectra of microdynamic impacts and the parameters of the technological process of crystal growth is primarily due to the fact that fast-flowing and slow-moving processes have to be compared. Adaptive evaluation algorithms and appropriate software have been developed to solve this problem.

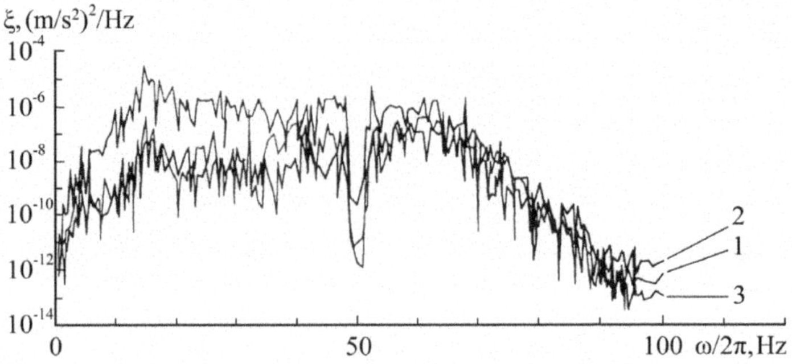

Fig. 4 Spectrograms of the power of microdynamic impacts during experiments

Experiments 1, 2 and 3 carried out on the modeling stand to measure vibration accelerations $\xi(t)$ (Fig. 4) from external and internal influencing factors in the frequency range from 0.01 to 100 Hz allowed us to build the following mathematical model:

$$\xi(t) = \sum_{k=1}^{r} A_k \cos(\omega_k t) + B_k \sin(\omega_k t) + \varepsilon(t), \quad t \in [0, T] \tag{1}$$

where T is the duration of the experiment, $\{A_k, B_k, \omega_k\}$ are the parameters of the technological process, $\varepsilon(t)$ is an ergodic high-frequency stationary random process with a spectrum $f(\omega)$.

A number of algorithms are available for calculating the model (1), i.e. measuring frequencies $\{\omega_k\}$, amplitudes $\{A_k, B_k\}$ and the spectrum $f(\omega)$ [12]. All available algorithms have one significant drawback: for their functioning, a priori information about the function $f(\omega)$ is necessary, for example, its smoothness class must be assumed to be known. Therefore, in the case of an unknown smoothness class, the convergence rate of classical methods may be very low.

Adaptive algorithms are known in the statistical literature [13–15] that do not use a priori information and, however, have the best convergence rate at $T \to \infty$, i.e. accuracy. The proposed method is a modification of the algorithms [13–15] adapted for this task.

Introduce two spectral processes into consideration:

$$\tau_1(\omega) = T^{-1} \int_0^T \xi(s) \cos(\omega s) ds, \quad \tau_2(\omega) = T^{-1} \int_0^T \xi(s) \sin(\omega s) ds,$$

and their envelope: $\tau(\omega) = \left\{[\tau_1(\omega)]^2 + [\tau_2(\omega)]^2\right\}^{0,5}$. By the ergodic theorem,

$$\lim_{T\to\infty} T^{-1} \int_0^T \varepsilon(s) \exp(i\omega s) ds.$$

Therefore, when $T \gg 1$ $T \gg 1$

$$T^{-1}\int_0^T \xi(s)\exp(i\omega s)ds \approx T^{-1}\sum_{k=1}^{r}[A_k\cos(\omega_k s) + B_k\sin(\omega_k s)]\exp(i\omega s)ds.$$

Calculating the integrals, we find that the spectral functions have limits

$$\lim_{T\to\infty}\tau_1(\omega) = 0, \quad \omega \neq \omega_k, \quad \lim_{T\to\infty}\tau_1(\omega) = A_k/2, \quad \omega = \omega_k, \tag{2}$$

and similarly

$$\lim_{T\to\infty}\tau_2(\omega) = 0, \quad \omega \neq \omega_k, \quad \lim_{T\to\infty}\tau_2(\omega) = B_k/2, \quad \omega = \omega_k. \tag{3}$$

Therefore

$$\lim_{T\to\infty}\tau(\omega) = 0, \quad \omega \neq \omega_k, \quad \lim_{T\to\infty}\tau(\omega) = \left(A_k^2 + B_k^2\right)^{0.5}/2, \quad \omega = \omega_k, \tag{4}$$

The following simple method of approximate determination of amplitudes and frequencies follows from the relations (2)–(4) $\{\omega_k\}$, $\{A_k\}$, $\{B_k\}$:

$$\omega_k^{(T)} = \arg\max_{\omega\in[\omega_0,\omega_1]} \tau(\omega), \quad A_k^{(T)} = \tau_1\left(\omega^{(T)}\right), \quad B_k^{(T)} = \tau_1\left(\omega^{(T)}\right),$$

where ω_0, ω_1 is the range of frequencies of practical interest in the problem under consideration. For example, $[\omega_0, \omega_1] = [0.01; 100]$ Hz.

In the practical calculation of the functions $\tau_1(\omega)$, $\tau_2(\omega)$, $\tau(\omega)$ the integrals are replaced by sums

$$\tau_1(\omega) \approx \tau_1^{(T)}(\omega) = \Delta T^{-1}\sum_{k=1}^{N}\cos(\omega t_k)\xi(t_k),$$

$$\tau_2(\omega) \approx \tau_2^{(T)}(\omega) = \Delta T^{-1}\sum_{k=1}^{N}\sin(\omega t_k)\xi(t_k),$$

$$\tau(\omega) \approx \tau^{(T)} = \left\{\left[\tau_1^{(T)}(\omega)\right]^2 + \left[\tau_2^{(T)}(\omega)\right]^2\right\}^{0,5},$$

where $t_k = \Delta k$, $N = T/\Delta$, Δ is the time step. To eliminate undesirable side maxima, the time steps Δ must be chosen so that $\Delta \ll 2\pi/\omega_1$.

The measurement of the spectrum $f(\lambda)$ is proposed to be carried out using the following adaptive algorithm. Consider the process:

$$\zeta(t) = \xi(t) - \sum_{k=1}^{r}\left\{A_k^{(T)}\cos\left[\omega_k^{(T)}t\right] + B_k^{(T)}\sin\left[\omega_k^{(T)}t\right]\right\},$$

Which is obtained by removing relatively low-frequency vibrations from the process $\xi(t)$: we will further refer to it as cleaning the process $\xi(t)$. Frequencies $\{\omega_k\}$, however, can affect the growth and quality of crystals and, therefore, may need damping or amplification.

Because of the initial process $\xi(t)$ are subtracted low-frequency oscillations, in the first approximation can be considered the process $\zeta(t)$ is stationary, as well as (in first approximation) assessment $\left\{A_k^{(T)}, B_k^{(T)}, \omega_k^{(T)}\right\}$ Gaussian, and process is Gaussian. More accurate calculations show that the unsteadiness of the process $\zeta(t)$ occurs only at the selected frequencies $\omega_k^{(T)}$, ω_k. In practice, the signal $\zeta(t)$ is recorded with a certain quantization step in time Δt, so that a time-discrete process (sequence) $\zeta_k = \zeta(\Delta t \times k)$ is available for processing.

To measure the spectral density of the sequence $\{\zeta_k\}, k = 1, 2, \ldots, n$, we decompose the spectrum $f(\lambda)$ into a Fourier series:

$$f(\lambda) = (2\pi)^{-1}\sum_{k=0}^{\infty} c_k\cos(k\lambda),$$

where $c_k = cov(\zeta_m, \zeta_{m+k})$ is the covariance function of the sequence $\{\zeta_k\}$. Let's say

$$a_n = n^{-1}\sum_{k=1}^{n}\zeta_k,\quad c_k(n) = n^{-1}\sum_{m=1}^{n-k}\zeta_m\zeta_{m+k} - a_n^2,\quad \psi(N) = \psi(n, N)\sum_{j=N}^{2N} c_k^2(n),$$

$$N_0 = N_0(n) = \arg\min_{N\in[2,n]} \psi(N).$$

As an adaptive estimate of the spectrum $f(\lambda)$, an estimate of the form is proposed

$$f_n(\lambda) = (2\pi)^{-1}\sum_{j=1}^{N_0} c_j(n)\cos(k\lambda),$$

This is a modification of the estimates [13–15]. The accuracy of the approximation $f \approx f_n$ and the confidence intervals for f are the same as in [13–15].

The measurement of mutual spectra provides meaningful information about the degree of dependence of various random processes at different frequencies. In particular, the dependence of crystal growth parameters on microdynamic influences is of interest for the problem under consideration.

Let's consider the developed adaptive algorithm for measuring mutual spectral density

$$f_{\xi,\eta}(\lambda) = \mathrm{g}(\lambda) = (2\pi)^{-1}\sum_{j=0}^{\infty} d_j \cos(\mathrm{k}\lambda)$$

of two stationary coupled processes $\eta_k, \xi_k, k = 1, 2, \ldots, n$, where $\eta_k = \eta(k, \Delta t), \xi_k = \xi(k, \Delta t)$ are microdynamic influences. Previously (if necessary) it is necessary to clean up the processes in question, as described earlier.

The proposed adaptive algorithm has not been considered in the world literature before.

Introduce the notations

$$a(n) = \mathrm{n}^{-1}\sum_{k=1}^{n} \xi_k, \quad b(\mathrm{n}) = \mathrm{n}^{-1}\sum_{k=1}^{n} \eta_k,$$

$$d_j = cov(\xi_k, \eta_{k+j}), \quad d_j(n) = \mathrm{n}^{-1}\sum_{k=1}^{n-j} \xi_k \eta_{k+j} - a(n)b(n),$$

$$\gamma(\mathrm{N}) = \gamma(\mathrm{n}, \mathrm{N}) = \sum_{j=N}^{2N} d_k^2(\mathrm{n}), \quad N_0 = N_0(n) = \arg\min_{N \in [2,n]} \gamma(\mathrm{N}).$$

As an adaptive estimate of the mutual spectrum $g(\lambda)$, we propose an estimate of the form

$$\mathrm{g}(\lambda) = (2\pi)^{-1}\sum_{j=1}^{N_0} d_j \cos(\mathrm{k}\lambda)$$

The properties of the adaptive estimate $g_n(\lambda)$ are completely identical to the properties of the previous estimates $f_n(\lambda)$.

A set of programs has been developed and debugged according to the proposed algorithm. The programs are verified on simulation models and during seismic measurements of industrial facilities in which the number of measurements is quite large (n = 2048). Thus, Fig. 5 shows adaptive (a, b) and classical (c, d) spectral estimates obtained for the same object at different points in time (a, b—spectrum estimates obtained by the proposed methods; c, d—classical spectral window methods [12]). The peaks of oscillations (Fig. 5b, d) are the result of poor-quality measurements. These fluctuations can forget the true frequency, resulting in reduced informativeness. Figure 5a, c shows that the proposed method is devoid of these disadvantages. The presented spectrograms clearly demonstrate the advantages of adaptive estimation, since it is free from high-frequency fluctuations caused by defects in nuclear estimates.

The proposed method of adaptive assessment allows, after measuring the frequencies of microdynamic influences affecting the parameters of technological processes, to issue recommendations for their suppression or enhancement using active, semi-active, passive or combined systems, most fully presented in the works [16]. When

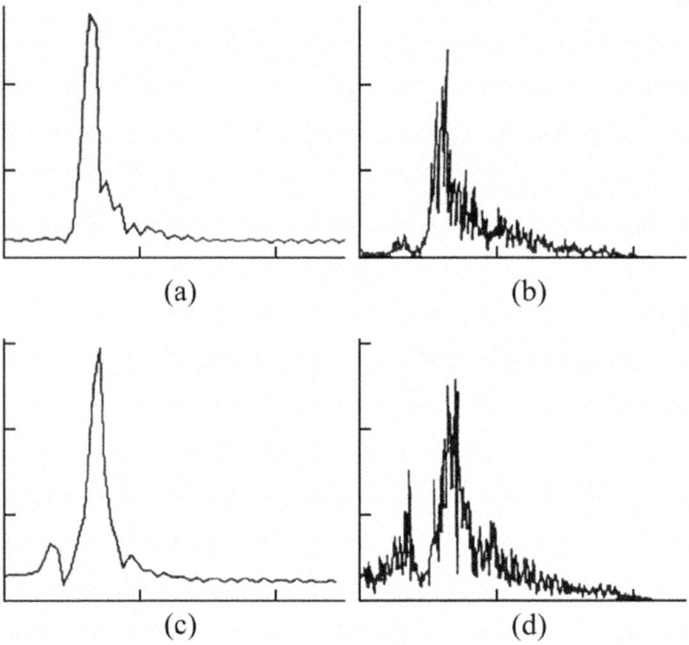

Fig. 5 Adaptive (**a**, **b**) and classical (**c**, **d**) estimates of seismic measurement spectra of the same industrial facility at different time points

organizing measurements in real time, this method can be used in algorithms for controlling active vibration isolation platforms.

3 Conclusion

It is noted that to date, no one in the world has managed to obtain crystals that, in terms of their combined properties, would surpass the best terrestrial analogues.

The necessity of creating modern space information technology systems for monitoring and controlling thermophysical processes with elements of artificial intelligence, made in the form of complex SEMS, is shown.

It is shown that the proposed assessment method allows, after measuring the frequencies of microdynamic influences affecting the parameters of technological processes, to make recommendations on their suppression or enhancement using active, semi-active, passive or combined systems. When organizing measurements in real time, this method can be used in algorithms for controlling active vibration isolation platforms.

The proposed method is universal and can be used not only in crystal growing processes, but also in other thermophysical processes occurring in microgravity.

The advantages of using the developed adaptive algorithms for evaluating the mutual influence of microdynamic action spectra and process parameters are shown.

References

1. Astronauts on the ISS grow crystals of solid solutions—News—Roscosmos State Corporation. https://www.roscosmos.ru/39657/ (in Russian). Accessed 07 Feb 2024
2. Redwire Sells Space-Manufactured Optical Crystal. https://www.photonics.com/Articles/Redwire_Sells_Space-manufactured_Optical_Crystal/a68155. Accessed 07 Feb 2024
3. Scientific program of the "Photon-M" spacecraft—News—Roscosmos State Corporation. https://www.roscosmos.ru/20614/ (in Russian). Accessed 07 Feb 2024
4. The "Photon-M No. 4" spacecraft. https://www.samspace.ru/products/satellites_of_scientific_purpose/ka_foton_m_4/ (in Russian). Accessed 07 Feb 2024
5. Strelov, V.I., Kuranova, I.P., Zakharov, B.G., Voloshin, A.E.: Crystallization in space: results and prospects. Crystallogr. Rep. **59**, 781–806 (2014)
6. The Russian orbital station. https://www.energia.ru/ru/spaceactivities/ros-station.html (in Russian). Accessed 07 Feb 2024
7. Ozernyh, I., Podturkin, A., Drakov, A., Makarenkov, A., SHishulin, V., Kolotovkin, A.: Kompleks kontrolya i upravleniya nauchno tekhnologicheskimi eksperimentami v kosmose [The complex of control and management of scientific and technological experiments in space]. Sovremennye tekhnologii avtomatizacii **4**, 42–47 (in Russian)
8. Gorodeckij, A.E., Tarasova, I.L., Kurbanov, V.G., Agapov, V.A.: Matematicheskaya model' sistemy avtomaticheskogo upravleniya modulya SEMS [Mathematical model of the SMS module automatic control system]. Informacionno-upravlyayushchie sistemy **3**, 40–45 (in Russian)
9. Sayapin, S.N., Artemenko, Y.N.: intelligence system for active vibration isolation and pointing of ultrahigh-precision large space structures in real time. In: Gorodetskiy, A. (ed.) Smart Electromechanical Systems. Studies in Systems, Decision and Control, vol. 49. Springer, Cham (2016). https://doi.org/10.1007/978-3-319-27547-5_10
10. Il'chenko, V.I.: Ul'trazvukovoj metod kontrolya krivizny i skorosti dvizheniya fronta kristallizacii [Ultrasonic method for monitoring the curvature and velocity of the crystallization front]. V mezhvedomstvennom nauchnom sb. "Dielektriki i poluprovodniki". Vypusk 9. Izd. "Vishcha shkola", Kiev, pp. 67–70 (1976) (in Russian)
11. YAcyk, S.I., Glotov, E.B.: Ustanovka dlya ul'trazvukovogo issledovaniya fronta kristallizacii v zharoprochnyh splavah [Installation for ultrasonic examination of the crystallization front in heat-resistant alloys]. Zavodskaya laboratoriya **40**, 192–193 (1974) (in Russian)
12. Andersen, R.S.: Statisticheskij analiz vremennyh ryadov [Statistical Analysis of Time Series]. Mir, Moscow (1977) (in Russian)
13. Bobrov, P.B., Ostrovskij, E.I.: Adaptivnoe ocenivanie regressii, plotnosti i spektra [Adaptive estimation of regression, density and spectrum]. Zavodskaya laboratoriya **5**, 55–57 (1996) (in Russian)
14. Ostrovskij, E.I.: Universal'noe adaptivnoe ocenivanie v trekh klassicheskih zadachah neparametricheskoj statistiki i bol'shie ukloneniya v Banahovom prostranstve [Universal adaptive estimation in three classical problems of nonparametric statistics and large deviations in Banach space]. Aktual'nye problemy sovremennoj matematiki **3**. Novosibirsk, NII MIOO, pp. 142–146 (1997) (in Russian)
15. Bobrov, P.B.: Metody adaptivnogo neparametricheskogo ocenivaniya v zadachah nadezhnosti i bezopasnosti YAEU [Methods of adaptive nonparametric estimation in problems of nuclear power plant reliability and safety]. Diss. kand. fiz.-mat. nauk, Obninsk, OIATE (in Russian) (1996)
16. Shi, H.T., Abubakar, M., Bai, X.T., Luo, Z.: Vibration isolation methods in spacecraft: a review of current techniques. Adv. Space Res. 1–31. https://doi.org/10.1016/j.asr.2024.01.020

Model Predictive Control and Genetic Algorithms for Optimization of Continuous Stirred Tank Reactors

Ali Deeb, Vladimir N. Khokhlovskiy, and Viacheslav P. Shkodyrev

Abstract *Problem statement*: The Continuous Stirred Tank Reactor is a classic example of a chemical reactor with nonlinear dynamics that is extensively studied and modeled in process control literature. However it is often considered using linearizations and other simplifications of a model of the object of control. *Purpose of research*: The material presents a sophisticated approach to controlling systems with inherently nonlinear dynamics (the Continuous Stirred Tank Reactor), through the advanced integration of a Genetic Algorithm (GA) with Model Predictive Control in order to avoid linearizations and other simplifications. While the application of Genetic Algorithm within the realm of Model Predictive Control is not novel and has been explored across a range of applications, the research under consideration stands out by offering a unique perspective: we optimize a consistent set of control parameters across the entire control horizon, as opposed to the more conventional approach of adjusting control inputs incrementally over the prediction horizon or merely optimizing for the immediate subsequent step. *Results:* The proposed innovative methodology is underpinned by a carefully crafted cost function, augmented with a discount factor that effectively addresses the limitations associated with optimizing solely for immediate future steps. By employing our approach, the process of determining control inputs is significantly simplified, which, in turn, markedly improves computational efficiency. The experimental findings robustly confirm that the proposed GA-enhanced Model Predictive Control strategy adeptly navigates the Continuous Stirred Tank Reactor towards the desired operational state, achieving superior performance over traditional control methods. *Practical significance*: The study underscores the adaptability and applicability of Genetic Algorithm within the

A. Deeb · V. N. Khokhlovskiy (✉) · V. P. Shkodyrev
SPbPU—Peter the Great St. Petersburg Polytechnic University, Saint-Petersburg, Russia
e-mail: hohlovskij_vn@spbstu.ru

A. Deeb
e-mail: dib_a@spbstu.ru

V. P. Shkodyrev
e-mail: shkodyrev@spbstu.ru

I. L. Tarasova and B. A. Kulik (eds.), *Smart Electromechanical Systems*, Studies in Systems, Decision and Control 544, https://doi.org/10.1007/978-3-031-64277-7_14

framework of Model Predictive Control with superior efficacy in managing intricate control scenarios, thus presenting a sophisticated solution in situations where conventional methods do not suffice.

Keywords Model predictive control · Genetic algorithms · Nonlinear dynamic systems · Continuous stirred tank reactor · Advanced control

1 Introduction

The objective of control engineering is to manipulate a dynamic system to attain a specific state or behavior. Key in this field are feedback controllers, which include classical, predictive, and repetitive types. Classical controllers like PID, bang-bang, and state controllers react to deviations based on past and current states [1]. Predictive controllers forecast future states using a system model [2], while repetitive controllers optimize future cycles based on past cycle behavior [3]. Despite the widespread use of PID controllers in industry, feedback control's efficacy is limited by system dynamics and model accuracy. Technical actuator constraints often necessitate heuristic, complex solutions. Advanced methods like sliding mode or backstepping controllers are available but complex [4]. While classic controls are deemed sufficient for most control problems [5, 6], recent insights suggest Model Predictive Control (MPC) is versatile enough for almost all scenarios, including those previously challenging due to limited knowledge or feasibility [2, 7]. Machine learning offers "intelligent" controllers that outperform adaptive controls in handling system nonlinearities and uncertainties [8, 9].

This study explores Genetic Algorithm (GA) for controlling the Continuous Stirred Tank Reactor (CSTR) without linearizations or other simplifications. Our contributions include a novel GA-based MPC strategy, the ability to control with the system's nonlinear model, and outperforming existing strategies in CSTR control.

The paper is structured as follows: problem formulation in Sect. 2, proposed algorithm in Sect. 3, experimental procedures and outcomes in Sect. 4, and conclusions and limitations in Sect. 5.

2 System Description

The Continuous Stirred Tank Reactor (CSTR) is a classic example of a chemical reactor that is extensively studied and modeled in process control literature. It is often controlled using advanced control strategies like Model Predictive Control (MPC) due to its nonlinear dynamics and the critical nature of maintaining specific output parameters. In the typical formulation of an MPC for a CSTR, the nonlinear system dynamics are represented by a set of differential equations that describe how the reactor's temperature and concentration change over time. These equations take into account factors such as the coolant flow rate, reaction kinetics, and heat transfer phenomena. The control objective is usually to track a desired setpoint for the output concentration while satisfying operational constraints and optimizing a cost function. This often involves solving an optimization problem at each control interval to determine the optimal coolant flow rate.

A modern model is proposed in [10] which integrates a first-principles model with a data-driven model to predict certain unknown dynamics of the process, resulting in higher accuracy than first-principles model. It is improved with the use of a deep neural network to provide stability. Contrary to this way we research the traditional MPC. The regular model of CSTR is explained in [11] where the authors developed a tube-based robust nonlinear model predictive control. In order to overcome disadvantages related to stochastic MPC authors in [12] consider a distributionally robust optimization method which reduces computational complexity. Nonetheless the proposed scheme is based on methods of linearization to deal with the inherent nonlinearity of CSTR. The approach was also discussed by Berberich et al. where they developed linear tracking MPC for nonlinear systems both model based [13] and the data-driven case [14]. Their strategy was to build a linearized dynamics at the current state as a prediction model. Then use these linearized models in MPC.

$$[\dot{x}_1, \dot{x}_2] = \left[\frac{T_s}{\theta}(1 - x_1) - T_s k_1 e^{\frac{-M}{x_2}}, \frac{T_s}{\theta}(x_f - x_2) + T_s k_1 e^{\frac{-M}{x_2}} - T_s \alpha u(x_2 - x_c)\right]$$

The states x_1 and x_2 are the temperature and the concentration, respectively, and the control input u is the coolant flow rate. These dynamics are obtained from [13]. We also use the same parameters with $T_s = 0.2, \theta = 20, k = 300, M = 5, x_f = 0.3947$, $x_c = 0.3816$, and $\alpha = 0.117$. Our control goal is tracking of the output setpoint $y_{ref} = 0.6519$ for the concentration while satisfying the input constraints $u_k \in U$ (where U is an interval from 0 to 1.2 with the open right edge) for $k \in N$.

3 Adapting Genetic Algorithm for Model Predictive Control

In the field of control systems, leveraging Genetic Algorithms (GA) as a Model Predictive Control (MPC) strategy presents a promising avenue for optimizing control parameters, particularly in managing systems with complex nonlinear dynamics. This approach entails an iterative search with GA to find control actions that minimize a specific cost function, usually framed as a weighted sum of squared differences between the system's actual output and the desired output.

The cost function aimed for minimization is articulated as:

$$J = \sum_{i=0}^{N-1} \gamma^i \left(y(i) - y_{ref}\right)^2$$

In this equation, $y(i)$ denotes the system output at step i, y_{ref} is the reference output, γ represents a damping factor within the range of 0 to 1, and N indicates the control horizon.

Our approach differentiates itself from prior applications of GA in MPC, such as the work by Vermuyten et al. [15], where GA was used to generate different time series of control inputs over the prediction horizon, which was time consuming and not suitable for a real-time implementation. Similarly, Wang et al. [16] employed MPC-GA for path tracking in autonomous vehicles, focusing on optimizing control values for the immediate next step. However, optimizing solely for the next step can lead to suboptimal outcomes in dynamic control systems, especially in complex nonlinear environments where allowing oscillations or overshooting before achieving stability can be advantageous.

In contrast, our methodology utilizes GA to optimize a single set of control parameters across the entire horizon. This strategy simplifies the control input process, reducing computational complexity, and is augmented by our specially formulated objective function with a discount factor. This adaptation, validated through experimental results, ensures that the system converges to the desired state efficiently, especially in scenarios where a broader horizon perspective is essential.

Algorithm 1 Genetic algorithm for MPC strategy

1. Procedure **GAMPC**
2. Initialize population with random control parameters
3. While termination condition not met do
4. Evaluate fitness using the system model and cost function J
5. Select the most suitable control parameters for reproduction
6. Crossover selected parameters to generate new candidates
7. Apply mutation to new candidates at a predetermined probability
8. Substitute the less suitable candidates with new ones
9. Return Optimal control parameters

4 Experimental Procedures and Outcomes

This research entailed conducting experiments on a personal computing system, specifically a machine equipped with Ubuntu 22.04, an Intel i5-12 processor, and 32 GB of RAM. Notably, no Graphics Processing Unit (GPU) was utilized in these experiments.

The test simulations were developed using Python, with the following Genetic Algorithm (GA) parameters: a population size of 10, 3 offsprings per generation, and a total of 100 generations, and finally a damping factor $\gamma = 0.9$. With these specified parameters, the system successfully identified a solution in under 0.1 s, demonstrating its appropriateness for real-time implementation in scenarios where $T_s = 0.2$.

Figure 1 shows plots with modeled data from [13] and compares behavior between linearization-based one-step MPC (solid line), linearization-based n-step

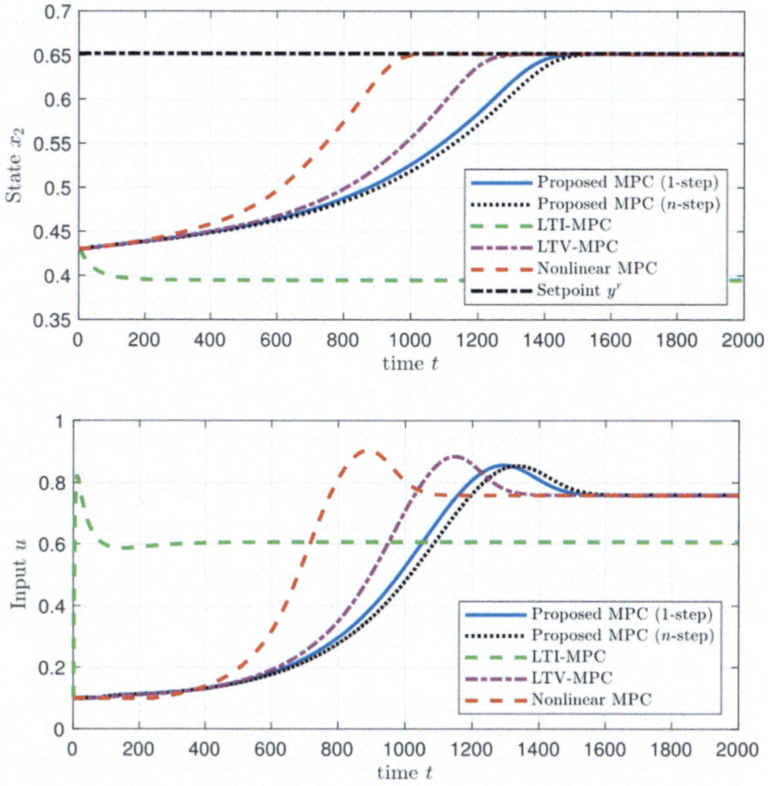

Fig. 1 Illustration of state components x_2 and the input component u of the system in

MPC (dotted line), LTI-MPC (dashed line), LTV-MPC (dash-dotted line), and nonlinear MPC (dashed line). The best achieved result to come to the stable state is about 1000 s for nonlinear MPC.

Figure 2 presents the outcomes of our proposed GAMPC strategy. Comparing to Fig. 1 displaying the system's response to the strategies proposed in [13] and [14], it is evident that our suggested approach significantly expedites the system's convergence, achieving it in less than 130 s, as opposed to the 1000 s reported in existing literature.

5 Discussion and Future Work

The research conducted in this study highlights the potential of Genetic Algorithms (GA) in advancing control engineering, particularly for systems like the Continuous Stirred Tank Reactor (CSTR). By leveraging a GA-based MPC strategy, we achieved effective control without the need for linearizations or dynamic simplifications, a

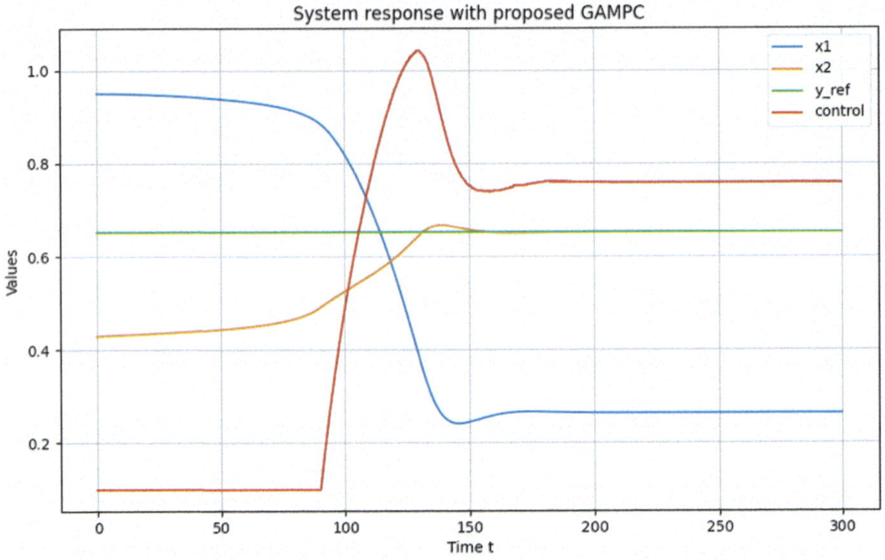

Fig. 2 Depiction of state components x_1 and x_2 along with the input component u of the system, employing our proposed GAMPC strategy

common limitation in traditional control strategies. This approach not only maintained the integrity of the system's nonlinear model but also demonstrated improved performance over conventional methods in terms of response time and adaptability. The results underscore the viability of GA in control applications, especially for systems with complex dynamics where traditional methods fall short.

While the GA-based MPC strategy shows promising results, there are inherent limitations. The primary constraint is the computational demand, as the GA search requires significant processing power. In our experiments, the computations were conducted on a standard CPU without a GPU, which, while sufficient for the CSTR system, might not scale effectively for more complex systems. This limitation suggests the need for more robust computational resources or optimized algorithms for larger-scale applications. Future research could explore the integration of parallel computing or the development of more efficient GA variants to address this challenge. Additionally, the real-world applicability of this approach needs to be tested in different industrial scenarios to validate its adaptability, scalability, and information interoperability beyond theoretical and controlled environments.

Acknowledgements The work is sponsored by the Russian Science Foundation, Grant 23-29-0051 of Jan.13, 2023.

References

1. Ang, H.K., Chong, G., Li, Y.: PID control system analysis, design, and technology. IEEE Trans. Control. Syst. Technol. **13**(4). 559–576 (2005)
2. Rawlings, J.B.: Tutorial overview of model predictive control. IEEE Control. Syst. Mag. **20**(3): 38–52 (2000)
3. Mayne, D.Q., Kerrigan, E.C., Van Wyk, E.J., Falugi, P.: Tube-based robust nonlinear model predictive control. Int. J. Robust Nonlinear Control. **21**(11), 1341–1353 (2011)
4. Vermuyten, E., Van den Zegel, B., Wolfs, V., Meert, P., Willems, P.: Real-time flood control by means of an improved MPC-GA algorithm and a fast conceptual river model for the Demer Basin in Belgium. In: Proceedings of the 6th International Conference on Flood Management, São Paulo, Brazil, pp. 16–18 (2014)
5. Wang, M., Chen, J., Yang, H., Wu, X., Ye, L.: Path tracking method based on model predictive control and genetic algorithm for autonomous vehicle. Math. Probl. Eng. (2022)
6. Hillerström, G., Walgama, K: Repetitive control theory and applications-a survey. IFAC Proc. **29**(1), 1446–1451 (1996)
7. Berberich, J., Köhler, J., Müller, M. A., Allgöwer, F.: Linear tracking MPC for nonlinear systems—part II: the data-driven case. IEEE Trans. Autom. Control. **67**(9), 4406–4421 (2022)
8. Banzhaf, W., Machado, P., Zhang, M.: Handbook of Evolutionary Machine Learning (2023)
9. Richalet, J.: Industrial applications of model based predictive control. Autom. **29**(5), 1251–1274 (1993)
10. Bangi, M.S.F., Kwon J.S.I.: Deep hybrid model-based predictive control with guarantees on domain of applicability. AIChE J. **69**(5), e18012 (2023). https://doi.org/10.1002/aic.18012
11. Dracopoulos, D.C.: Evolutionary Learning Algorithms for Neural Adaptive Control. Springer (2013)
12. Garcia, C.E., Morari, M.: Internal model control. A unifying review and some new results. Ind. Eng. Chem. Process. Des. Dev. **21**(2), 308–323 (1982)
13. Zhong, Z., del Rio-Chanona, E.A., Petsagkourakis, P.: Tube-based distributionally robust model predictive control for nonlinear process systems via linearization. Comput. Chem. Eng. **170**, 108112 (2023). https://doi.org/10.1016/j.compchemeng.2022.108112
14. Zinober, A.S.I., Owens, D.H.: Nonlinear and Adaptive Control. Springer Science & Business Media, p. 281 (2002)
15. Schwenzer, M., Ay, M., Bergs, T., Abel, D.: Review on model predictive control: an engineering perspective. Int. J. Adv. Manuf. Technol. **117**(5–6):1327–1349 (2021)
16. Berberich, J., Köhler, J., Müller, M.A., Allgöwer, F.: Linear tracking MPC for nonlinear systems—part I: the model-based case. IEEE Trans. Autom. Control. **67**(9), 4390–4405 (2022)

Decision-Making When Assessing Financing Risks Using the Logical and Linguistic Method

Andrey E. Gorodetskiy, Irina L. Tarasova, and Anna R. Krasavtseva

Abstract *Problem statement*: Decision-making when assessing the risks of lending to individuals is associated with the uncertainty of the influence of many factors that cannot be accurately described mathematically. This leads to situations that reduce the likelihood of assessing the risks of non-refund. Among the various methods for assessing the risk of non-repayment of funds, one can highlight the scenario approach, which uses methods of fuzzy set theory to calculate the values of membership functions. However, the problem of ranking a set of non-repayment risk scenarios, taking into account the mental properties of experts and borrowers, has not been sufficiently studied. *The purpose*: Assessing the risks of financing and lending to individuals, taking into account the borrower's temperament and the preferences of experts based on the logical–linguistic method. Development of algorithms for calculating the risk assessment of lending to the analyzed individual and determining its rating. *Results*: Based on the proposed risk assessment method, an algorithm for calculating the risk assessment of non-repayment of funds has been developed. Using computer modeling, it is shown that when the opinions of experts vary, setting the significance coefficients of borrower indicators at approximately 30%, individual experts can give a worse risk forecast than the average assessment of several experts. Therefore, when assessing the risk of loan non-repayment of funds, it is advisable to use at least 5 experts and rank the borrower according to the average score. *Practical significance*: The research results can be used to create a methodology for assessing the risks lending to individuals.

Keywords Lending to individuals · Risk assessments · Expert · Borrower · Scoring models and coefficients · Preference coefficients · Significance coefficients · The standard · Logical–linguistic classification · Rating · Decision-making

---

A. E. Gorodetskiy · I. L. Tarasova (✉)
Institute for Problems in Mechanical Engineering of the Russian Academy of Sciences, Saint-Petersburg, Russia
e-mail: g17265@yandex.ru

A. R. Krasavtseva
Saint-Petersburg, Russia

I. L. Tarasova and B. A. Kulik (eds.), *Smart Electromechanical Systems*, Studies in Systems, Decision and Control 544, https://doi.org/10.1007/978-3-031-64277-7_15

1 Introduction

Making a decision on lending to individuals is associated with the uncertainty of the influence of many factors that cannot be accurately mathematically described, which leads to situations that reduce the likelihood of assessing the risks of non-repayment of funds. Therefore, in practice, risks are usually associated with uncertainty [1]. However, Knight [2] proposed to distinguish the concepts of "risk" and "uncertainty". When assessing risks, actions are considered that may entail several scenarios for the development of events with a known distribution of their probabilities. When the distribution is unknown, situations are regarded as uncertainties [3], which can be described by membership functions in the formulation of Zadeh [4].

Qualitative and quantitative approaches to risk assessment are used in risk assessment [1, 5, 6]. The qualitative approach is characterized by solving the following tasks: analysis of the subject area, identification of risky stages of work, identification of the causes of risk; identification of emerging types of risks; forecasting possible consequences that may occur when implementing a risk decision [7]. Based on the qualitative analysis, the main indicators of the borrower are formulated. When conducting a qualitative analysis, an expert method is usually used. During its implementation, experts evaluate each type (indicator) of risk, after that all expert assessments are collected and processed using methods of system analysis [8]. The results of the qualitative analysis serve as the basis for further quantitative analysis of the investment project. The quantitative approach determines the numerical value (scoring coefficients) of the risk indicators identified in the qualitative analysis [9, 10], the assessment of the risks of financing borrowers and their ranking [11]. In this case, methods of probability theory, statistics, operations research, as well as logical-probabilistic and logical–linguistic classification methods can be used [12, 13]. These methods usually complement each other in an in-depth risk analysis.

Since expert opinions are used in risk analysis, the results of the analysis may be influenced to one degree or another by the mental characteristics of experts, as well as borrowers when lending to individuals, especially their temperament, which has not been sufficiently studied. Therefore, despite the existing technical difficulties of accounting for the psyche, it is advisable to rank risks according to the average opinion of experts [11], and when evaluating loans to individuals, additionally introduce indicators of their temperament [14]. By temperament, people differ from each other from birth [15], namely, the strength of the processes of excitation and inhibition; the balance of these processes; mobility (variability) of the processes of excitation and inhibition. Moreover, the properties of temperament manifest themselves depending on the situation and specific conditions. People with different temperaments can act completely differently in identical situations. When identifying temperament indicators, the expert should take into account the recommendations of psychologists and the characteristic properties of the borrower's temperament that affect his creditworthiness.

2 Credit Risk Assessment Models

Currently, the main models used to assess the risks of lending to individuals are: credit scoring (Durand model) [16], probity model [17] and decision tree [18]. Neural networks are also widely used [19]. In addition, back in the early 60s, methods were proposed that added psychological and sociological concepts to the applied criteria of risk assessment models. In addition to the traditional characteristics of borrowers, it was proposed to consider psychological characteristics that influenced the degree of credit risk [20].

In a dialogue with a borrower to assess his creditworthiness, it is advisable to pre-evaluate his belonging to one or another type of temperament. Modern concepts of temperament [15] try to combine different ways of classifying temperament. The analysis of various methods of classifying human temperament carried out in [15] showed that experts can use the following:

- Constitutional, based on the analysis of human images [21];
- Interactive, based on the analysis of human responses to the questionnaire items [22–25];
- Behavioral, based on the analysis of the characteristics of human behavior in various situations [26].

Since in practice situations of fluctuations in a person's temperament often arise when various life situations arise, it is difficult to take them into account when evaluating a person and using any one of the listed approaches. Therefore, it is necessary to determine the degree of belonging of a person to a certain type of temperament after using all three classification methods. Then, if all three types of classification give the same result (for example, sanguine), then the degree of belonging to the identified type of classification will be equal to $\mu(\text{sanguine}) = \mu(c) = 1$ in terms of linguistic variables [4]. If all three give different results, then we get, for example:

$$\mu(\text{sanguine}) = \mu(\text{choleric}) = \mu(\text{phlegmatic}) \\ = \mu(c) = \mu(x) = \mu(f) = 1/3.$$

It is most likely that two types of classification will give the same result, and the third one is excellent. Then, for example, we get: $\mu(c) = 2/3$ and $\mu(x) = 1/3$.

In the case when all three types of classification give different results, the task of assessing temperament becomes uncertain. One of the promising approaches to solving such a problem may be the use of logical-probabilistic and logical–linguistic methods of analyzing the solution selection environment, which allow correcting the classification results. At the same time, it seems advisable to expand the list of temperament types with correction of the degree of belonging to a particular type, based on a logical-probabilistic and logical–linguistic analysis of the results of interaction in various situations. However, there should not be infinitely many correction rules.

After identifying the borrower's temperament, the expert can set the scoring coefficients of the borrower's temperament indicators. At the same time, it is advisable to take into account some recommendations of psychologists [15]. For example:

- When assessing choleric, it must be remembered that not everything that a choleric says is really true. It's just that his speech is ahead of his thoughts, depending on the conditions and specific emotions at the moment. Under other conditions, he may say the opposite things. Do not put pressure on the choleric, especially in matters of a personal nature.
- When evaluating a sanguine person, keep in mind that communication is the air for a sanguine person and it seems that it is impossible to quarrel with him, but this is not the case. Due to the superficiality of judgments, sanguine people often draw wrong conclusions, but it is difficult for a sanguine person to admit his mistakes, he does not know how to do this. The face is an instruction for a sanguine person. Everything is really written on the face. Be observant, and the sanguine facial expressions will tell you everything: what a person thinks, what is pleasant to him, and what is unpleasant and more.
- When evaluating a melancholic, avoid loud and harsh sounds, screams and do not resort to a negative assessment of a melancholic. Preference should be given to warm, sincere conversations. In a detailed job that requires focus and attention, a melancholic is an excellent candidate. Melancholic have a well-developed intuition, they rarely make mistakes.
- When evaluating a phlegmatic person, remember that you can get information from a phlegmatic person only through communication, but you need to pull everything out with ticks. Phlegmatic people are conservatives, so you should not require them to quickly adopt something new. Do not expect efficiency and speed in other matters. When communicating with a phlegmatic person, clear, practical, understandable speech is required. It is recommended to avoid imagery and slang when communicating with a phlegmatic person.

When analyzing a borrower, in addition to the "temperament" indicator, it is proposed to evaluate him according to other criteria:

1. Age (over 75 years, from 61 to 75 years, from 18 to 24 years, from 40 to 60 years, from 25 to 39 years);
2. Field of activity (IT, manufacturing, education, science, medicine, commerce, arts etc.);
3. Education (incomplete secondary, secondary, higher, postgraduate school, academic degree);
4. Marital status (marriage to a foreigner, registered marriage, unregistered marriage, divorced);
5. Availability of dependents: (more than three dependents, three dependents, two dependents, one dependent, no dependents;
6. Housing conditions: renting an apartment, house in a village, owning an apartment, owning a country house, owning a country house and owning an apartment;

7. Availability of a car (no car, economy class for more than 5 years, economy class for up to 5 years, business class for more than 5 years, business class for up to 5 years);
8. Work experience (no, up to 2 years, from 2 to 5 years, from 5 to 10 years, more than 10 years;
9. Place of residence (rural area, resort, city with a population of less than 1 million, a city with a population of 1 million to 5 million, a population of more than 5 million;
10. Profession (pensioner, worker (employee), farmer (entrepreneur), head of a unit, head of an enterprise (institution).

The proposed borrower indicators and the corresponding scoring points are determined by experts. Therefore, before assessing lending risks, experts need to agree on their opinions about the standards and come to the same assessments. Experts can display their preferences in the values of preference coefficients for each borrower indicator. However, the spread in preference coefficients should not be too large. For example, in computer modeling, the spread of the influence of expert opinions should not exceed 30%.

3 Ranking of Borrower's Indicators by Risk Degree

In a scenario approach that uses fuzzy set theory methods to calculate the values of membership functions, it is advisable to rank the risks of lending to an individual using credit scoring by comparing them with etalon indicators from a database (DB) based on the logical–linguistic classification method. In this regard, at the very beginning, a database of the borrower's etalon indicators is created, which is a set of etalon strings and corresponding scoring, which are set by experts. The following are the borrower's etalon indicators and their corresponding data:

1. Age:
   - Over 75 years old (etalon string A_{11} = /10000/ and scoring $C_{11} = 1$);
   - From 61 to 75 years old (etalon string A_{12} = /01000/ and scoring $C_{12} = 2$);
   - From 18 to 24 years old (etalon string A_{13} = /00100/ and scoring $C_{13} = 3$);
   - from 40 to 60 years old (etalon string A_{14} = /00010/ and scoring $C_{14} = 4$),
   - From 25 до 39 years old (etalon string A_{15} = /00001/ and scoring $C_{15} = 5$);
2. Temperament:
   - Indeterminate (etalon string A_{21} = /10000/ and scoring $C_{21} = 1$),
   - Phlegmatic (etalon string A_{22} = /01000/ and scoring $C_{22} = 2$),
   - Choleric (etalon string A_{23} = /00100/ and scoring $C_{23} = 3$),
   - Sanguine (etalon string A_{24} = /00010/ and scoring $C_{24} = 4$),
   - Melancholic (etalon string A_{15} = /00001/ and scoring $C_{25} = 5$);
3. Field of activity:

- Student (etalon string A_{31} = /10000/ and scoring $C_{31} = 1$);
- Science, education, medicine (etalon string A_{32} = /01000/ and scoring $C_{32} = 2$);
- Industry and services (etalon string A_{33} = /00100/ and scoring $C_{33} = 3$);
- Military (etalon string A_{34} = /00010/ and scoring $C_{34} = 4$);
- Civil servant (etalon string A_{35} = /00001/ and scoring $C_{35} = 5$).

4. Education:

   - Incomplete secondary (etalon string: A_{41} = /10000/ scoring $C_{41} = 1$),
   - Secondary (etalon string A_{42} = /01000/ and scoring $C_{42} = 2$),
   - Higher (etalon string: A_{43} = /00100/ and scoring $C_{43} = 3$),
   - Graduate school (etalon string A_{44} = /00010/ and scoring $C_{44} = 4$),
   - Academic degree (etalon string A_{45} = /00001/ and scoring $C_{45} = 5$);

5. Family status:

   - Marriage with a foreigner (etalon string A_{51} = /10000/ and scoring $C_{51} = 1$),
   - Civil marriage (etalon string A_{52} = /01000/ and scoring $C_{52} = 2$),
   - Divorced (etalon string A_{53} = /00100/ and scoring $C_{53} = 3$),
   - Not married (etalon string эталонная строка A_{54} = /00010/ and scoring $C_{54} = 4$),
   - Married (etalon string A_{55} = /00001/ and scoring $C_{55} = 5$);

6. Availability of dependents:

   - More than three dependents (etalon string A_{61} = /10000/ and scoring $C_{61} = 1$),
   - Three dependents (etalon string A_{62} = /01000/ and scoring $C_{62} = 2$),
   - Two dependents (etalon string A_{63} = /00100/ and scoring $C_{63} = 3$),
   - One dependent (etalon string A_{64} = /00010/ and scoring $C_{64} = 4$),
   - No dependents (etalon string A_{65} = /00001/ and scoring $C_{65} = 5$);

7. Housing conditions:

   - Apartment rental (etalon string A_{71} = /10000/ and scoring $C_{71} = 1$),
   - House in the village (etalon string A_{72} = /01000/ and scoring $C_{72} = 2$),
   - Owned apartment (etalon string A_{73} = /00100/ and scoring $C_{73} = 3$),
   - Own cottage (etalon string A_{74} = /00010/ and scoring $C_{74} = 4$),
   - Owned cottage and apartment (etalon string A_{75} = /00001/ and scoring $C_{75} = 5$);

8. Availability of a car

   - No car (etalon string A_{81} = /10000/ and scoring $C_{81} = 1$),
   - Economy class for more than 5 years (etalon string A_{82} = /01000/ and scoring $C_{82} = 2$),
   - Economy class up to 5 years (etalon string A_{83} = /00100/ and scoring $C_{83} = 3$),

- Business class for more than 5 years (etalon string A_{84} = /00010/ and scoring $C_{84} = 4$),
- Business class up to 5 years (etalon string A_{85} = /00001/ and scoring $C_{85} = 5$);

9. Work experience:

   - No (etalon string A_{91} = /10000/ and scoring $C_{91} = 1$),
   - Up to 2 years (etalon string A_{92} = /01000/ and scoring $C_{92} = 2$),
   - From 2 to 5 years (etalon string A_{93} = /00100/ and scoring $C_{93} = 3$),
   - From 5 to 10 years (etalon string A_{94} = /00010/ and scoring $C_{94} = 4$),
   - More than 10 years (etalon string A_{95} = /00001/ and scoring $C_{95} = 5$);

10. Place of residence:

    - Rural area (etalon string A_{101} = /10000/ and scoring $C_{101} = 1$),
    - Spa (etalon string A_{102} = /01000/ and scoring $C_{102} = 2$),
    - A city of less than 1 million (etalon string A_{103} = /00100/ and scoring $C_{103} = 3$),
    - A city of 1 million to 5 million (etalon string A_{104} = /00010/ and scoring $C_{104} = 4$),
    - A city of more than 5 million (etalon string A_{105} = /00001/ and scoring $C_{105} = 5$);

11. Profession:

    - Pensioner (etalon string A_{111} = /10000/ and scoring $C_{111} = 1$),
    - Worker (employee) (etalon string A_{112} = /01000/ and scoring $C_{112} = 2$),
    - Farmer (entrepreneur) (etalon string A_{113} = /00100/ and scoring $C_{113} = 3$),
    - Head of the department (etalon string A_{114} = /00010/ and scoring $C_{114} = 4$),
    - Head of the enterprise (institution) (etalon string A_{115} = /00001/ and scoring балл $C_{115} = 5$).

Next, during a survey or dialogue with the borrower, information is obtained on the proposed indicators. To obtain information about the borrower's temperament, the popular Eysenck test [26] is usually used, which is easily implemented by using an expert system and a monitor dialog box or, which is somewhat more difficult, a speech dialogue. In this case, the borrower's performance is ranked according to the following algorithm.

1. Introduction of etalon credit benchmarks $i = 1, 2, \ldots, N$;
2. For each i-th creditworthiness indicator, significance coefficients k_i are assigned, for example: $k_1 = 1.2$, $k_2 = 1.5$, $k_3 = 1.3$, $k_4 = 1.1$, $k_5 = 1.4$, $k_6 = 1.6$, $k_7 = 1.8$, $k_8 = 1.7$, $k_9 = 1.0$, $k_{10} = 1.9$, $k_{11} = 2.0$;
3. For each i-th creditworthiness indicator, preference coefficients are assigned for men S_i^1 and women S_i^2, for example: $S_1^1 = 1$ and $S_1^2 = 0.95$, $S_2^1 = 1$ and $S_2^2 = 1.1$, $S_3^1 = 1$ and $S_3^2 = 0.98$, $S_4^1 = 1$ and $S_4^2 = 0.95$, $S_5^1 = 1$ and $S_5^2 = 1.1$, $S_6^1 = 1$ and $S_6^2 = 1.1$, $S_7^1 = 1$ and $S_7^2 = 1$; $S_8^1 = 1$ and $S_8^2 = 0.95$, $S_9^1 = 1$ and $S_9^2 = 1.0$, $S_{10}^1 = 1$ and $S_{10}^2 = 1.0$, $S_{11}^1 = 1$ and $S_{11}^2 = 0.95$;

4. For each i-th indicator, depending on the type of its value $j = 1, 2, \ldots, M$, a scoring C_{ij} is assigned, for example, for the indicator "age" ($i = 1$) with a value of the type "from 25 to 39 years "($j = 5$) - scoring $C_{15} = 5$;
5. Calculation of the maximum risk assessment for lending to an individual:

$$\max\{O_R^1\} = \frac{1}{N}\sum_{i=1}^{N} \max\{r_{ij}^1\},$$

where, k_m and S_m—maximum values of coefficients;
6. Introduction of risk assessments of non-repayment of the loan, for example, Table 1.
7. Formation of string with the values of the membership functions μ_{ij} of the attributes of the indicators of the analyzed borrower, for example: A_{5a} = / 0 0.1 0.9 0 0/;
8. Assignment of all string of values of membership functions of the analyzed borrower to the etalon ones for each indicator using a method of image classification similar to the patent [13];
9. Assigning scoring values to the analyzed borrower for all types of indicators of the selected etalon borrower, for example A_{5a} corresponds to A_{53} and therefore $C_{5a} = C_{53} = 3$;
10. Calculation of the risk assessment of lending to the analyzed individual, for example, for a man:

$$O_{\text{Ra}}^1 = \frac{1}{N}\sum_{i=1}^{N} \{r_{ia}^1\},$$

where: $\{r_{ia}^1\} = (k_i * S_i^1)/(C_{ia})$
11. Determining the rating of the analyzed individual and making a decision on lending in accordance with Table 1.

Since the assignment of numerical values k_i, S_i^1 or S_i^2 and C_{ij} for the etalons depends on the experts, to take into account their opinions it is advisable to conduct a similar assessment for several experts and choose the average.

Table 1 Loan non-repayment risk assessments

Rank	Borrower risk assessment O_{Ra}	Characteristic
1	From 0 to (1/4) max $\{O_R^1\}$	Excellent borrower
2	From (1/4) max $\{O_R^1\}$ to (1/2)max $\{O_R^1\}$	Good borrower, but some minor comments
3	From (1/2) max $\{O_R^1\}$ to (3/4)max $\{O_R^1\}$	Satisfactory borrower and additional information required
4	From (3/4) max $\{O_R^1\}$ to max $\{O_R^1\}$	Unsatisfactory borrower

4 Analysis of the Influence of Expert Opinions on Risk Assessments of Lending to Individuals and Determining Their Ratings

Analysis of the influence of expert opinions on risk assessments of lending to individuals and determining their ratings was carried out using computer modeling. At the same time, 5 experts were involved, 5 male borrowers and 5 female borrowers. The significance coefficients kip and preferences S_{ip}^1, S_{ip}^2 were selected by each of the 5 experts from the following strings:

$$k = /1\ 1.1\ 1.2\ 1.3\ 1.4\ 1.5\ 1.6\ 1.7\ 1.8\ 1.9\ 2.0/;$$

$$S = /0.95\ 0.96\ 0.97\ 0.98\ 0.99\ 1.00\ 1.01\ 1.02\ 1.03\ 1.04\ 1.05/.$$

Scoring for the values of all *ij*-th etalons indicators of creditworthiness were established from the set $C = /1\ 2\ 3\ 4\ 5/$. In particular, for all *i*-th indicators there were the following: $C_{i1} = 1$, $C_{i2} = 2$, $C_{i2} = 2$, $C_{i3} = 3$, $C_{i4} = 4$, $C_{i5} = 5$.

The values of the membership functions of scoring *ij*-th strings of all *q*-th analyzed borrowers $A_{ia}{}^q$ with the format as in the etalon strings A_{ij} were chosen randomly using a random number generator. Moreover, in each $A_{ia}{}^q$ line there could be no more than two values of membership functions that are not equal to zero.

When calculating the maximum risk assessment for lending to an individual, the following was used: $k_m = 1.5$, $S_m = 1$, $C_{ij} = 1$. Therefore: $\max\{O_R{}^1\} = 1.5$, (1/4) $\max\{O_R{}^1\} = 0.375$; (1/2) $\max\{O_R{}^I\} = 0.75$, (3/4) $\max\{O_R{}^1\} = 0.825$.

The modeling results were as follows:

Coefficients of the first expert:

- String of significance coefficients: $\boldsymbol{k} = /1.2\ 1.6\ 1.3\ 1.7\ 1.7\ 1.1\ 1.8\ 1.9\ 1.4\ 1.8\ 1.5/$;
- Preference coefficient string for a man: $\boldsymbol{S}^1 = /1\ 1\ 1\ 1\ 1\ 1\ 1\ 1\ 1\ 1\ 1/$;
- Preference coefficient string for a woman: $\boldsymbol{S}^2 = /1.01\ 0.99\ 0.96\ 0.99\ 0.95\ 0.95\ 1.03\ 1\ 1.03\ 1.01\ 1/$;

Coefficients of the second expert:

- String of significance coefficients: $\boldsymbol{k} = /1.2\ 1.3\ 1\ 1.6\ 1.3\ 1.9\ 1.3\ 1.5\ 1.1\ 1.3\ 1.1/$;
- Preference coefficient string for a man: $\boldsymbol{S}^1 = /1\ 1\ 1\ 1\ 1\ 1\ 1\ 1\ 1\ 1\ 1/$;
- Preference coefficient string for a woman: $\boldsymbol{S}^2 = /0.98\ 0.98\ 1\ 1.01\ 0.98\ 0.98\ 0.95\ 0.95\ 0.97\ 0.97\ 1.01/$;

Coefficients of the third expert:

- String of significance coefficients: $\boldsymbol{k} = /1\ 1.2\ 1.6\ 1.7\ 1.4\ 1.6\ 1.8\ 1\ 1.2\ 1\ 1.1/$;
- Preference coefficient string for a man: $\boldsymbol{S}^1 = /1\ 1\ 1\ 1\ 1\ 1\ 1\ 1\ 1\ 1\ 1/$;
- Preference coefficient string for a woman: $\boldsymbol{S}^2 = /1\ 0.99\ 0.98\ 0.97\ 0.99\ 1.02\ 1.02\ 1\ 1.01\ 1\ 0.95/$;

Coefficients of the fourth expert:

- String of significance coefficients: $\boldsymbol{k} = /1.1\ 1.2\ 1.1\ 1.2\ 1.3\ 1.4\ 1\ 1.5\ 1.2\ 1.1\ 1.1/$;
- Preference coefficient string for a man: $\boldsymbol{S}^1 = /1\ 1\ 1\ 1\ 1\ 1\ 1\ 1\ 1\ 1\ 1/$;
- Preference coefficient string for a woman: $\boldsymbol{S}^2 = /1\ 0.99\ 0.98\ 0.97\ 0.99\ 1.02\ 1.02\ 1\ 1.01\ 1\ 0.95/$;

Coefficients of the fifth expert:

- String of significance coefficients: $\boldsymbol{k} = /1.4\ 1.2\ 1.1\ 1.2\ 1.9\ 1.1\ 1.4\ 1.1\ 1.3\ 1\ 1.3/$;
- Preference coefficient string for a man: $\boldsymbol{S}^1 = /1\ 1\ 1\ 1\ 1\ 1\ 1\ 1\ 1\ 1\ 1/$;
- Preference coefficient string for a woman: $\boldsymbol{S}^2 = /1\ 0.99\ 0.98\ 0.97\ 0.99\ 1.02\ 1.02\ 1\ 1.01\ 1\ 0.95/$;

Characteristics of the first borrower, man and woman:

- strings of values of membership functions of scoring indicators: $\boldsymbol{A}_{1a}{}^1 = /0\ 0\ 0\ 0.2\ 0.8/$, $\boldsymbol{A}_{2a}{}^1 = /0\ 0\ 1\ 0\ 0/$, $\boldsymbol{A}_{3a}{}^1 = /0\ 1\ 0\ 0\ 0/$, $\boldsymbol{A}_{4a}{}^1 = /0\ 0\ 0\ 0\ 1/$, $\boldsymbol{A}_{5a}{}^1 = /0\ 0\ 0\ 0\ 1/$, $\boldsymbol{A}_{6a}{}^1 = /0\ 0.2\ 0.8\ 0\ 0/$, $\boldsymbol{A}_{7a}{}^1 = /0\ 1\ 0\ 0\ 0/$, $\boldsymbol{A}_{8a}{}^1 = /0\ 0\ 0.7\ 0.3\ 0/$, $\boldsymbol{A}_{9a}{}^1 = /0\ 0\ 0\ 0.1\ 0.9/$, $\boldsymbol{A}_{10a}{}^1 = /0\ 0\ 0\ 0.2\ 0.8/$, $\boldsymbol{A}_{11a}{}^1 = /0\ 0\ 0\ 0\ 1/$;
- Scoring string after the borrower classification: $\boldsymbol{C}^1 = /5\ 3\ 2\ 5\ 5\ 3\ 2\ 3\ 5\ 5\ 5/$.

Characteristics of the second borrower, man and woman:

- strings of values of membership functions of scoring indicators: $\boldsymbol{A}_{1a}{}^2 = /0\ 0\ 0\ 0.2\ 0.8/$, $\boldsymbol{A}_{2a}{}^2 = /0.2\ 0\ 0.8\ 0\ 0/$, $\boldsymbol{A}_{3a}{}^2 = /0\ 0\ 0\ 0\ 1/$, $\boldsymbol{A}_{4a}{}^2 = /0\ 0\ 0\ 0\ 1/$, $\boldsymbol{A}_{5a}{}^2 = /0\ 0\ 0\ 1\ 0/$, $\boldsymbol{A}_{6a}{}^2 = /0\ 0\ 0\ 0.9\ 0.1/$, $\boldsymbol{A}_{7a}{}^2 = /1\ 0\ 0\ 0\ 0\ 1/$, $\boldsymbol{A}_{8a}{}^1 = /0\ 0\ 0.8\ 0.2\ 0/$, $\boldsymbol{A}_{9a}{}^1 = /0\ 0\ 0\ 0.1\ 0.9/$, $\boldsymbol{A}_{10a}{}^1 = /0\ 0\ 0\ 0.1\ 0.9/$, $\boldsymbol{A}_{11a}{}^1 = /0\ 0\ 0\ 0\ 1/$;
- scoring string after the borrower classification: $\boldsymbol{C}^2 = /5\ 3\ 5\ 5\ 4\ 4\ 5\ 3\ 5\ 5\ 5/$.

Characteristics of the third borrower, man and woman:

- strings of values of membership functions of scoring indicators: $\boldsymbol{A}_{1a}{}^3 = /0\ 0\ 0\ 0.2\ 0.8/$, $\boldsymbol{A}_{2a}{}^3 = /0.2\ 0\ 0.8\ 0\ 0/$, $\boldsymbol{A}_{3a}{}^3 = /0\ 0\ 0\ 0\ 1/$, $\boldsymbol{A}_{4a}{}^3 = /1\ 0\ 0\ 0\ 0/$, $\boldsymbol{A}_{5a}{}^3 = /1\ 0\ 0\ 0\ 0/$, $\boldsymbol{A}_{6a}{}^3 = /0\ 0.9\ 0.1\ 0\ 0/$, $\boldsymbol{A}_{7a}{}^3 = /1\ 0\ 0\ 0\ 0/$; $\boldsymbol{A}_{8a}{}^3 = /0\ 0\ 0\ 0.2\ 0.8/$, $\boldsymbol{A}_{9a}{}^3 = /0.9\ 0.1\ 0\ 0\ 0/$, $\boldsymbol{A}_{10a}{}^3 = /0\ 0\ 0\ 0.2\ 0.8/$, $\boldsymbol{A}_{11a}{}^3 = /0\ 0\ 0\ 0\ 1/$;
- scoring string after the borrower classification: $\boldsymbol{C}^3 = /5\ 3\ 5\ 1\ 1\ 2\ 1\ 5\ 1\ 5\ 5/$.

Characteristics of the fourth borrower, man and woman:

- strings of values of membership functions of scoring indicators: $\boldsymbol{A}_{1a}{}^4 = /0\ 0\ 0\ 0.2\ 0.8/$, $\boldsymbol{A}_{2a}{}^4 = /0.1\ 0\ 0\ 0\ 0.9/$, $\boldsymbol{A}_{3a}{}^4 = /0\ 0\ 0\ 0\ 1/$, $\boldsymbol{A}_{4a}{}^4 = /0\ 0\ 0\ 0\ 1/$, $\boldsymbol{A}_{5a}{}^4 = /0\ 0\ 0\ 0\ 1/$, $\boldsymbol{A}_{6a}{}^4 = /0\ 0.1\ 0.9\ 0\ 0/$, $\boldsymbol{A}_{7a}{}^4 = /0\ 0\ 0\ 0\ 1/$, $\boldsymbol{A}_{8a}{}^4 = /0.8\ 0.2\ 0\ 0\ 0/$, $\boldsymbol{A}_{9a}{}^4 = /0\ 0.2\ 0.8\ 0\ 0/$, $\boldsymbol{A}_{10a}{}^4 = /0\ 0\ 0\ 0.1\ 0.9/$, $\boldsymbol{A}_{11a}{}^4 = /0\ 0\ 0\ 1\ 0/$;
- scoring string after the borrower classification: $\boldsymbol{C}^4 = /5\ 5\ 5\ 5\ 5\ 3\ 5\ 1\ 3\ 5\ 4/$.

Characteristics of the fifth borrower, man and woman:

- strings of values of membership functions of scoring indicators: $\boldsymbol{A}_{1a}{}^5 = /0\ 0.1\ 0.9\ 0\ 0/$, $\boldsymbol{A}_{2a}{}^5 = /0.8\ 0\ 0\ 0\ 0.2/$, $\boldsymbol{A}_{3a}{}^5 = /0\ 0\ 0\ 0\ 1/$, $\boldsymbol{A}_{4a}{}^5 = /1\ 0\ 0\ 0\ 0/$, $\boldsymbol{A}_{5a}{}^5 = /0\ 0.2\ 0.8\ 0\ 0/$, $\boldsymbol{A}_{6a}{}^5 = /0\ 0\ 0\ 0.2\ 0.8/$, $\boldsymbol{A}_{7a}{}^5 = /0\ 0\ 0\ 0\ 1/$, $\boldsymbol{A}_{8a}{}^5 = /0\ 0\ 0\ 0.1\ 0{,}9/$, $\boldsymbol{A}_{9a}{}^5 = /0\ 0\ 0\ 0.2\ 0.8/$, $\boldsymbol{A}_{10a}{}^5 = /0\ 0\ 0\ 0.1\ 0.9/$, $\boldsymbol{A}_{11a}{}^5 = /1\ 0\ 0\ 0\ 0/$;

- scoring string after the borrower classification: $C^5 = /3\ 1\ 5\ 1\ 3\ 5\ 5\ 5\ 5\ 5\ 1/$.

The obtained results of computer modeling are shown in Figs. 1, 2, 3, 4 and 5.

As can be seen from Figs. 1, 2, 3, 4 and 5 assessments of the same borrower by various experts, the spread of the significance coefficients of which reached 30%, and the preference coefficients—10%, basically fit into the same rank. However, there are also deviations, for example (see Fig. 1) the first, second, third and fifth experts gave in accordance with the Table 1 to the first borrower rank 2 (Good borrower, but there are minor comments), and the fourth expert—rank 1 (Excellent borrower). Therefore, it is advisable to rank borrowers by averaging the opinions of several

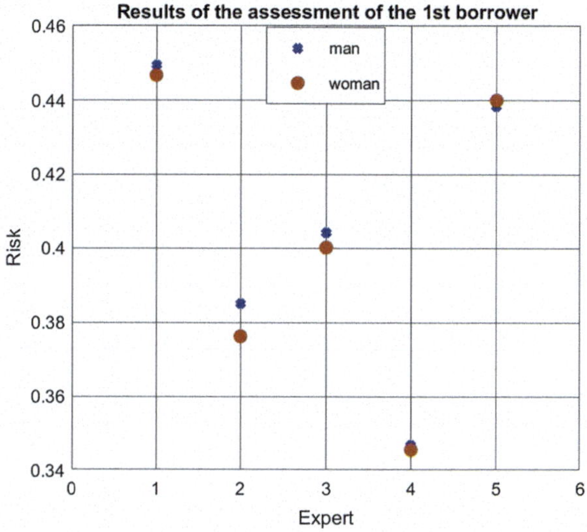

Fig. 1 Dependence of the risk of lending to the 1st borrower on the expert's opinion

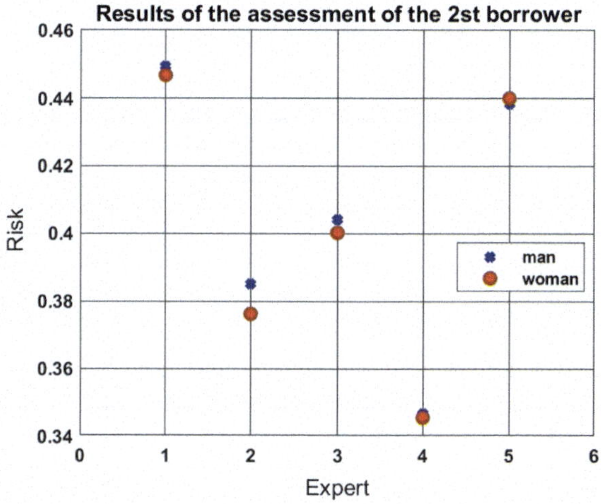

Fig. 2 Dependence of the risk of lending to the 2st borrower on the expert's opinion

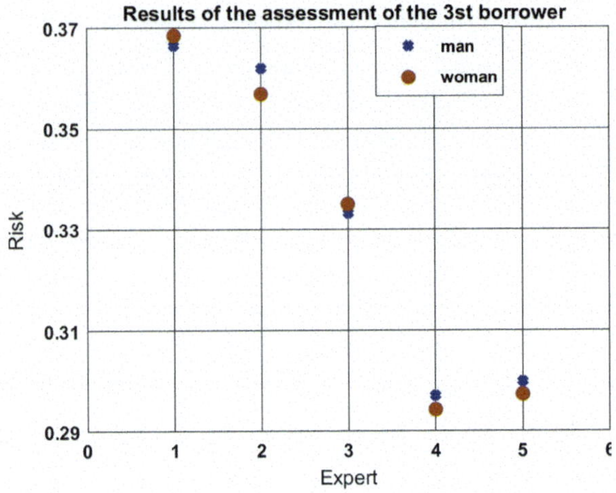

Fig. 3 Dependence of the risk of lending to the 3st borrower on the expert's opinion

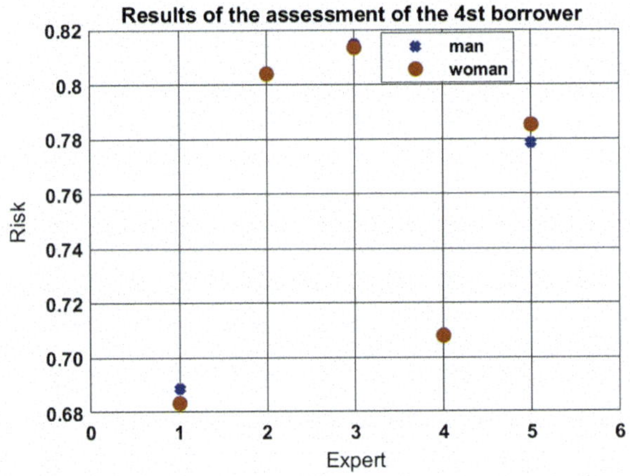

Fig. 4 Dependence of the risk of lending to the 4st borrower on the expert's opinion

experts. The differences in the assessments of man and woman borrowers in the given example were insignificant, which can be explained by the small spread of preference coefficients in a given set.

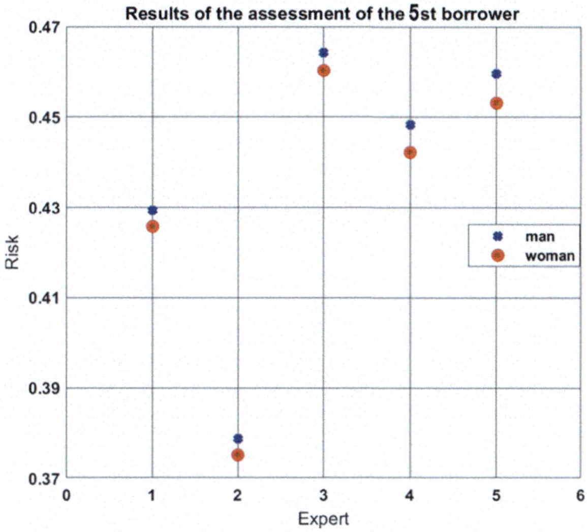

Fig. 5 Dependence of the risk of lending to the 5st borrower on the expert's opinion

5 Conclusion

When assessing the risks of lending to individuals, it is proposed to use a logical–linguistic method of risk assessment and an algorithm for ranking borrowers, which increases the accuracy of the assessment and the speed of ranking due to the possibility of simple computer implementation. At the same time, the mental characteristics of experts are reflected in the coefficients of significance and preferences that they select from the proposed sets, and the mental characteristics of borrowers are reflected in indicators of their temperament.

The results of computer modeling of risk assessments of loan non-repayment by various borrowers showed the dependence of the assessments on the coefficients established by various experts. Therefore, it is advisable to rank the risks of loan non-repayment due to differences in expert opinions by averaging the opinions of several experts when comparing borrowers' indicators with etalon data from the database.

The main part of the proposed algorithm for ranking borrowers is implemented computer-based in a registered computer program [27].

To evaluate the efficiency and stability of the algorithm, computer simulation can be used to generate hundreds of possible combinations of attributes (importance coefficients, preferences and scoring) and the values of their membership functions, taking into account their probability distribution [28].

Acknowledgements The work was carried out within the framework of the state assignment of the Ministry of Science and Higher Education of the Russian Federation for the Institute of Problems in Mechanical Engineering of Russian Academy of Sciences.

References

1. Popova, A.Y.: Risk assessment of an investment project. Sci. J. KubSAU. **19**, 1–26 (2006)
2. Knight, F.H.: Risk, uncertainty and profit. Transl. from English—M.: Delo p. 360 (2003)
3. Shtele, E.A., Guseva, M.A., Rudi, L.A.: Methodology for assessing the effectiveness of investment projects taking into account risks. SibADI Bull. **6**(52), 135–140 (2016)
4. Zadeh, L.A.: The Concept of a Linguistic Variable and its Application to Approximate Decision Making, p. 165. M.:Mir, (1976). (In Russian)
5. Moskvin, V.A.: Risks of Investment Projects, p. 320. M: Course: INFRA-M, (2016). (In Russian)
6. Reshetnyak, E.I.: Methods for assessing investment risks in business planning. Bus. Inform. **12**, 189–194 (2017). (In Russian)
7. Bender, M., Panz, S.: A common framework for identifying and categorizing risks: an application to the context of financial markets. J. Risk. Comput. Sci. (2021)
8. Tkachenko, A.L., Sheveleva, O.G.: Risk assessment in investment design. Omsk Sci. Bull. Ser. Soc. Story Modernity **4**(2), 140–145 (2019). (In Russian)
9. Borisova, O.V., Malykh, N.I., Oveshnikova, L.V.: Investments: Textbook and Workshop for Universities, 2nd ed., revised, p. 483. Additional M.: Yurayt, (2023). (In Russian)
10. Vedmed, I.Y.: Analysis of quantitative methods for assessing the risks of investment projects. Probabilistic Mathematical Models of Economic Processes: Sat. Scientific Articles of the XII Conference Russian Regions in the Focus of Change, pp. 52–61. Ekaterinburg, (2017). (In Russian)
11. Korolkova, E.M.: Risk management: project risk management: textbook. Manual for Students of Economic Specialties. Tambov: Publishing House of the Federal State Budgetary Educational Institution of Higher Professional Education TSTU, p. 160 (2013). (In Russian)
12. Krasavtseva, A.R., Gorodetskiy, A.E.: Logical-linguistic method for assessing the risk of specialized lending (using the example of project financing). Econ. Math. Methods **58**(4), 83–91 (2022). (In Russian)
13. Gorodetskiy, A.E., Kurbanov, V.G., Tarasova, I.L.: Method of Image Classification Patent RU No. 2756778 (2021)
14. Gorodetskiy, A.E., Tarasova, I.L., Kurbanov, V.G.: Classification of images in decision making in the central nervous system of SEMS. In: Smart Electromechanical Systems. Behavioral Decision Making. Studies in Systems, Decision and Control 352. Springer Nature Switzerland AG, (2021). https://doi.org/10.1007/978-3-030-68172-2_15
15. Klinga, L., König-Kersting, C., Trautmanna, S.T.: Investment preferences and risk perception: financial agents versus clients. J. Bank. Financ. T. **154**, 106489 (2023)
16. Chovdyrova, G.S., Klimenko, T.S.: Clinical psychology. General Part: Textbook. Benefit, p. 247. M.: UNITY-DANA, (2017). (In Russian)
17. Gorodetskiy, A.E., Tarasova, I.L.: Optimizing the movement of a robot router taking into account the driver's temperament. Probl. Artif. Intell. **1**(28), 18–30 (2023). (In Russian)
18. Ishina, I.V., Sazonova, M.N.: Scoring model for assessing credit risk. Audit Fin. Anal. **4**, 297–304 (2007). (In Russian)
19. Nivorozhkina, L.N., Ovcharova, L.N., Sinyavskaya, T.G.: Economic modeling of the risk of payments on consumer loans. Appl. Econometrics. **30**, 65–76 (2013). (In Russian)
20. Larin, S.: Using decision trees to assess the creditworthiness of individuals. BaseGroup/abs. Data Analysis Technologies. https://basegroup.ru/community/articles/solvency
21. Gorodetskiy, A.E., Tarasova, I.L.: Control and Neural Networks, p. 312. SPb.: Publishing House Polytechnic. University, (2005). (In Russian)
22. Berezhnoy, M.A., Maksimova, N.N.: Review of models for assessing the creditworthiness of individuals. Bull. AmSU **89**, 164–169 (2020). (In Russian)
23. Kretschmer, E.: Body Structure and Character, p. 327. Publishing house, Academic Project, (2015). (In Russian)
24. Greben' N.F.: Psychological Tests for Professionals, p. 496. We'll lie. School, Minsk (2007)

25. Rusalov, V.M.: Temperament Structure Questionnaire (TSQ), p. 50. IP AN SSSR, M. Mir, (1990). (In Russian)
26. Personality Questionnaire EPI (G. Eysenck method). Almanac of Psychological Tests, pp 217–224. M., (1995). (In Russian)
27. Collection of Psychological Tests. Part I: Manual. Comp. HER. Mironova, p. 155. Women's Institute ENVILA, Mn, (2005)
28. Mendelevich, V.D.: Clinical and Medical Psychology. Textbook Allowance, p. 432. Publishing house, MEDpress-inform, (2008)

Risk Analysis of Project Financing Using the Logical–linguistic Classification Method

Andrey E. Gorodetskiy, Anna R. Krasavtseva, and Irina L. Tarasova

Abstract *Problem statement*: The assessment of credit risks is predictive in nature, associated with the uncertainty of the influence of many factors that cannot be accurately described mathematically. This leads to a low probability of obtaining and repayment of loans for various facilities, lending including specialized. There are usually two main types of risk assessment methods—qualitative and quantitative. Based on these methods, it is possible to identify a scenario approach using the methods of fuzzy set theory to calculate the values of membership functions. However, the problem of ranking the scenarios for the implementation of the project for which a loan is requested has not been sufficiently studied. *The purpose*: Analysis of specialized exposures on the example of project financing using methods of logical–linguistic classification and ranking method. Building a database of criteria for credit requirements for project financing and developing an algorithm for calculating the risk assessment of lending to the analyzed project. *Results*: The developed algorithm makes it possible to increase the reliability and speed of evaluation of the analyzed project to be financed. According to this method, the possibility of non-repayment of the loan can be predictable. *Practical significance*: The results of the study can be used in the development of a computer program that allows you to speed up the risk analysis of credit projects.

Keywords Specialized lending · Risk assessment · Project financing · Creditworthiness level · Weighting factors · Significance coefficients · Etalons · Logical–linguistic classification · Rating · Decision-making

A. E. Gorodetskiy · I. L. Tarasova
Institute for Problems in Mechanical Engineering of the Russian Academy of Sciences, Saint-Petersburg, Russia

A. R. Krasavtseva (✉)
Saint-Petersburg, Russia
e-mail: krasavtseva.a@gmail.com

I. L. Tarasova and B. A. Kulik (eds.), *Smart Electromechanical Systems*, Studies in Systems, Decision and Control 544, https://doi.org/10.1007/978-3-031-64277-7_16

1 Introduction

The assessment of credit risks is predictive in nature, associated with the uncertainty of the influence of many factors that cannot be accurately mathematically described. This causes situations that reduce the likelihood of successful loan repayment.

There are usually two main types of risk assessment methods—qualitative and quantitative [1–3].

With a qualitative approach, possible types of risks of non-fulfillment of loan repayment obligations (risks of default) are determined, as well as areas of occurrence and sources of possible risks are determined [4]. It can serve as a basis for obtaining quantitative risk assessments.

The quantitative approach determines the numerical value of separate risks over the life of the loan [5, 6]. In this case, methods of probability theory, statistics, and operations research can be used. These methods usually complement each other in an in-depth risk analysis.

Among the qualitative methods of risk assessment, the following can be noted:

- Analysis of possible cases of non-fulfillment of obligations (default);
- The method of analogies,
- The method of expert assessments.

The subjective nature of such assessments is dampened by the well-developed organization of all stages of the examination and the use of quantitative methods, both in the organization of the examination and in evaluating the judgments of experts based on the processing of the results of their opinions. In general, the examination includes the following stages: the preliminary stage; the formation of a working group; the formation of an expert group; the preparation and conduct of a survey of experts; processing and analysis of the results of the survey of experts. The preliminary stage is one of the most important steps of the expertise.

Among the quantitative methods, the following are distinguished:

- Statistical analysis;
- Sensitivity analysis;
- Analysis of sustainability verification and scenarios,
- Imulation modeling.

To obtain an acceptable result of expert assessments, it is necessary to apply a complex procedure for organizing expert assessments [7], starting with the selection of the number and qualifications of experts. The results of the multi-stage procedure are processed using statistical and qualitative analysis methods. For a more complete risk analysis, options based on regression and correlation analysis are used, and logical-probabilistic approaches can be used to detail and analyze complex projects [8]. The decision tree method is additionally used for scenario analysis. These are network graphs, each branch of which represents probabilistic options for the development or status of the project [9].

2 Project Financing Within the Framework of Specialized Lending

Specialized lending includes various classes such as: project financing (*A*), object financing (*B*), commodity financing (*C*), financing of income-generating real estate (*D*), financing of estate with unstable price parameters (*E*). Let's consider one of the classes, namely project financing.

Project financing is the provision of credit products to clients for the purpose of financing or refinancing investment projects in a wide variety of industries, including financing the acquisition of "ready-made businesses." At the same time, the main source of repayment of financing is the cash flow generated during the implementation of the investment project.

As part of project financing, lending is carried out for the construction or reconstruction of large facilities for borrowers (for example, such as energy and transport infrastructure facilities), while the assets created during the implementation of the project are assets are pledged as collateral for the loan.

Models of project financing may differ depending on the specifics of project financing, the specific purpose of the project, as well as the type of agreement (contract) that is the basis for financing. But there are some general principles:

- Project financing is used to finance a relatively "isolated" project through a legal entity specializing in the implementation of this project (SPV);
- The source of return on invested funds is the profit from the investment project implementation (separated from the financial results of the activities of the project initiators); assets related specifically to the investment project can serve as additional collateral;
- The share of attracted capital in the total volume of project financing can reach 70–90%;
- Project financing usually deals with more or less well-known technologies, and projects are most often focused on the production of traditional goods and services.

Thus, for the bank in project financing there is an additional risk of non-return of invested funds due to the risks of the investment project. Such risks include: delay in commissioning of the facility, exceeding the construction budget, low quality of equipment and construction work, increase in prices for raw materials and unqualified management of the facility at the operation stage, etc.

3 Algorithm for Ranking Projects by Risk Level

Due to the limited statistical data when forecasting project financing risks, it is advisable to create a database of reference projects containing qualitative attributes and their expert quantitative assessments. These estimates are presented in the form of values of membership functions of attributes and values of coefficients of their significance, as proposed in the logical–linguistic method of classification [10].

Then, with a scenario approach using fuzzy set theory methods to calculate the values of membership functions, it is possible to rank project implementation scenarios by comparing them with etalon projects from the database [10]. At the same time, to assess the effectiveness and sustainability of scenarios, one can use computer simulation modeling, which generates hundreds of possible combinations of project attributes and the values of their membership functions, taking into account their probability distribution [11, 12]. Then, after analyzing the modeling results, it is possible to obtain a probabilistic distribution of possible lending results and give an integral assessment of the effectiveness of financing [13].

If, according to some criterion, the risk of project implementation is higher than 95%, then such a project is removed from consideration.

A project accepted for consideration is ranked according to the following algorithm.

1. Assignment of weight coefficients k_j for all j-th subclasses of credit requirements (project attributes) for the analyzed class A projects.
2. Assignment of risk degrees R_{jiq} for all jiq–th values of criteria types (attributes).
3. Assignment of risk assessment (level of project creditworthiness) (for example, Table 1 [14])
4. Formation of a database of etalon strings of attribute for all class A projects.
5. Formation strings of attribute of the analyzed project.
6. Assignment of the analyzed project to one of the etalon ones using an image classification method similar to patent [15].
7. Assignment of risk degree values for the analyzed project for all values of the types of criteria (attributes) of the selected reference project.
8. Calculation of the risk assessment of lending for the analyzed project.
9. Determining the rating of the analyzed project and making a decision.

Let's take a closer look at pp. 4–9 of the algorithm using the example of ranking a class A project.

Table 1 Ranks and corresponding financing risk assessments

Rank	Borrower risk assessment O_R	Characteristic
1	From 0 to 19	Excellent borrower
2	From 19 to 38	Good borrower, but some minor comments
3	From 38 to 57	Satisfactory borrower and additional information required
4	From 57 to 76	Unsatisfactory borrower

4 Formation of Etalon Strings of Attribute of Projects Class *A* (Project Financing)

Section I [16] provides criteria for credit requirements classified as class *A*, which are used to assess the risks of lending to project finance. These criteria are based on the recommendations of the Basel Committee on Banking Supervision [17].

From the credit requirements criteria, it is possible to generate etalon strings of attribute for each class, indicating the degree of risk for each attribute.

For example, for class *A* (project financing) we enter the following etalon strings:

A_{1ji}—Financial position. Weight coefficient $k_1 = 5$ (*i*, *j*—number of section and subsection of criteria for creditworthiness requirements from the classifier [16]);

A_{11i}—Market conditions. Significance coefficient $k_{11} = 5$;

A_{111} = /1000/—(Few competing suppliers or a significant and long-term advantage in location, cost price or technology. Demand is strong and growing). The degree of risk $R_{111} = 1$;

A_{112} = /0100/—(Few competing suppliers and (or) the location, cost price or technology is better than average, but this situation cannot last long. Demand is strong and stable). The degree of risk $R_{112} = 2$;

A_{113} = /0010/—(Project has no advantages in terms of location, cost price or technology. The demand is adequate and stable) Degree of risk $R_{113} = 3$;

A_{114} = /0001/—(Location, cost price or technology is below average. Demand is weak and declining). Degree of risk $R_{114} = 4$;

A_{115} = /0000/—(No information available). Degree of risk $R_{115} = 5$;

A_{12i}—(Financial indicators (including debt service coverage, ratio of borrowed and own funds, etc.) Significance coefficient $k_{12} = 4$;

A_{121} = /1000/—(High indicators of financial ratios, taking into account the possibility of adverse consequences in the form of losses of expected investment income, very reliable economic forecasts). Degree of risk $R_{121} = 1$;

A_{122} = /0100/—(High and acceptable indicators of financial ratios, taking into account the possibility of adverse consequences in the form of losses of expected investment income, reliable project economic forecasts). Degree of risk $R_{122} = 2$;

A_{123} = /0010/—(Standard indicators of financial ratios, taking into account the possibility of adverse consequences in the form of losses of expected investment income). Degree of risk $R_{123} = 3$;

A_{124} = /0001/—(Low indicators financial ratios, taking into account the possibility of adverse consequences in the form of loss of expected investment income). Degree of risk $R_{124} = 4$;

A_{125} = /0000/—(No information available). Degree of risk $R_{125} = 5$;

A_{13i} - Stress analysis. Significance coefficient $k_{13} = 3$;

A_{131} = /1000/—(Financial obligations under the project can be fulfilled in conditions of prolonged and serious problems in the economy or industry). Degree of risk $R_{131} = 1$;

A_{132} = /0100/—(The project's financial obligations may be met during normal economic or industry problems. The project may default only if economic conditions are severe). Degree of risk $R_{132} = 2$;

A_{133} = /0010/—(Project success is sensitive to stresses common to the business cycle, and the project may default during a normal downturn). Degree of risk $R_{133} = 3$;

A_{134} = /0001/—(The project may default if conditions do not improve soon). Degree of risk $R_{134} = 4$;

A_{135} = /0000/—(No information available). Degree of risk $R_{135} = 5$;

A_{14i}—Financial structure (Weighted average term to repayment of the loan, calculated using the current value of cash flows for this loan, compared with the weighted average term to completion of the project, calculated using the current value of cash flows for this project). Significance coefficient $k_{14} = 2$;

A_{141} = /1000/—(The weighted average time to completion of the project, calculated using the current value of cash flows for this project, significantly exceeds the weighted average time to repayment of the loan, calculated using the current value of cash flows for this loan). Degree of risk $R_{141} = 1$;

A_{142} = /0100/—(The weighted average term to completion of the project, calculated using the current value of cash flows for this project, exceeds the weighted average term to repayment of the loan, calculated using the current value of cash flows for this loan). Degree of risk $R_{142} = 2$;

A_{143} = /0010/—(The weighted average term to completion of the project, calculated using the current value of cash flows for this project, slightly exceeds the weighted average term to repayment of the loan, calculated using the current value of cash flows for this loan). Degree of risk $R_{143} = 3$;

A_{144} = /0001/—(The weighted average term to completion of a project, calculated using the present value of cash flows for a given project, may not exceed the weighted average term to maturity of the loan, calculated using the present value of cash flows for a given loan). Degree of risk $R_{144} = 4$;

A_{145} = /0000/—(No information available). Degree of risk $R_{145} = 5$;

A_{15i}—Repayment schedule. Significance coefficient $k_{15} = 1$;

A_{151} = /1000/—(Fully amortizing loan). Degree of risk $R_{151} = 1$;

A_{152} = /0100/—(Partially amortizing loan). Degree of risk $R_{152} = 2$;

A_{153} = /0010/—(Installment payments with limited lump sum repayment). Degree of risk $R_{153} = 3$;

A_{154} = /0001/—(Lump sum repayment or repayment in installments with a high proportion of lump sum repayment (bullet loan)). Degree of risk $R_{154} = 4$;

A_{155} = /0000/—(No information available). Degree of risk $R_{155} = 5$;

The following is a classification of sections and subsections of criteria for loan requirements related to the class of project financing [16]. Based on this classification, etalon attribute strings are generated for each subclass, taking into account the degree of risk for each attribute, as shown above.

A_{2ji}—Political and legal environment. Weight coefficient $k_2 = 4$;

A_{21i}—The possibility of losses or a reduction in expected profit due to changes in government policy, including the risk of restrictions on the transfer of capital, taking into account the type of project and risk mitigating factors). Significance coefficient $k_{21} = 6$;

A_{22i}—Risk of force majeure (war, unrest, etc.). Significance coefficient $k_{22} = 5$;

A_{23i}—Government support and strategic importance of the project. Significance coefficient $k_{23} = 4$;

A_{24i}—Stability of the legal and regulatory environment (risk of changes in legislation). Significance coefficient $k_{24} = 3$;

A_{25i}—Obtaining necessary permits in accordance with local legislation. Significance coefficient $k_{25} = 2$;

A_{26i}—Ensuring legal sanctions for the execution of contracts and the sale of collateral. Significance coefficient $k_{25} = 1$;

A_{3ji}—Characteristics of operations. Weight coefficient $k_3 = 1$;

A_{31i}—The risk of losses due to engineering errors in the development of the project, errors in technology. Significance coefficient $k_{31} = 6$;

A_{32i}—Construction risk. Significance coefficient $k_{32} = 5$;

$A_{321} = /1000/$—(All initial permits for construction have been received). Degree of risk $R_{321} = 1$;

A_{33i}—The type of construction contract. Significance coefficient $k_{33} = 5.5$;

A_{34i}—Guarantees of completion of construction. Significance coefficient $k_{34} = 4.5$;

A_{35i}—Work experience and financial situation of contractors on the example of the implementation of similar projects. Significance coefficient $k_{35} = 4$;

A_{36i}—Risk of possible losses due to non-fulfillment of the terms of contracts for operational and maintenance services. Significance coefficient $k_{36} = 3.5$;

A_{37i}—Work experience and financial position of the operator. Significance coefficient $k_{37} = 3$;

A_{38i}—The risk of selling products. The buyer's obligation to accept the goods or pay a penalty or a fixed price contract. Significance coefficient $k_{38} = 2.5$;

A_{39i}—The absence of the buyer's obligation to accept the goods or pay a penalty or a fixed price contract. Significance coefficient $k_{39} = 2$;

A_{310i}—The risk of supply. Suppliers' failure to fulfill their obligations due to the implementation of transport or price risks; financial position and experience of suppliers. Significance coefficient $k_{310} = 1.5$;

A_{311i}—The risks of losses due to insufficient material reserves (for example, the risks of insufficient reserves of natural resources in the field during its development). Significance coefficient $k_{311} = 1$;

A_{4ji}—The quality of the sponsor. Weight coefficient $k_4 = 3$;

A_{41i} – Sponsor's work experience, financial situation, experience in a given country or industry. Significance coefficient $k_{41} = 2$;

A_{42i}—Sponsorship, confirmed by equity participation, ownership conditions and incentives to contribute additional funds if necessary. Significance coefficient $k_{42} = 1$;

A_{5ji}—Provision. Weight coefficient $k_5 = 2$;

A_{51i}—Assignment of contracts and accounts. Significance coefficient $k_{51} = 5$;

A_{52i}—Pledge of assets taking into account their quality, value and liquidity. Significance coefficient $k_{52} = 4$;

A_{53i}—Creditor's control over cash flows (for example, receipt of funds, independent escrow accounts). Significance coefficient $k_{53} = 3$;

A_{54i}—Covenants under the agreement (mandatory advance payments, deferred payments, sequence of payments, restrictions on dividends). Significance coefficient $k_{54} = 2$;

A_{55i}—Availability of funds reserved for established needs (for example, debt service, extension and replacement, unforeseen events, and so on). Significance coefficient $k_{55} = 1$;

5 Formation of a Database of Attribute Strings of the Analyzed Project

For the analyzed project, it is not always possible to attribute the attribute value to one or another of the four types entered. For example, for the attribute A_{22i}—Risk of force majeure, it is not always possible to say for sure that A_{221} = /1000/—(High protection) or A_{222} = /0100/—(Acceptable protection). In this regard, for the analyzed project L, the type of this attribute is set by probability values, or by a membership function, for example A_{22L} = /0.7 0.3 0 0/. Tasking probability values involves the need for statistical processing of data, which may be insufficient or missing. Therefore, it is better to use the values of membership functions set by experts, or obtained on the basis of fuzzification of the available limited data [12, 18, 19].

Thus, after forming the attribute strings of the analyzed project, there will be a string database of the form $A_{jiL} = /\mu_1\ \mu_2\ \mu_3\ \mu_4\ /$. Moreover, only two values of μ_i may not be equal to 0.

For example, the following strings are generated for the analyzed project L from class A (project financing):

$$A_{11L} = /0.7\ 0.3\ 0\ 0/; A_{12L} = /0.3\ 0.7\ 0\ 0/; A_{13L} = /0\ 0.3\ 0.7\ 0/; A_{14L} = /0\ 0.7\ 0.3\ 0/;$$
$$A_{15L} = /0.6\ 0.4\ 0\ 0/; A_{21L} = /0\ 0.6\ 0.4\ 0/; A_{22L} = /0.6\ 0.4\ 0\ 0/; A_{23L} = /0.7\ 0.3\ 0\ 0/;$$
$$A_{24L} = /0.8\ 0.2\ 0\ 0/; A_{25L} = /0\ 0.7\ 0.3\ 0/; A_{26L} = /0\ 0.6\ 0.4\ 0/; A_{31L} = /0\ 0.7\ 0.3\ 0/;$$
$$A_{32L} = /0.3\ 0.7\ 0\ 0/; A_{33L} = /0\ 0\ 0.7\ 0.3/; A_{34L} = /0\ 0\ 0.6\ 0.4/; A_{35L} = /0\ 0\ 0.4\ 0.6/;$$
$$A_{36L} = /0.6\ 0.4\ 0\ 0/; A_{37L} = /0\ 0.3\ 0.7\ 0/; A_{38L} = /0\ 0.7\ 0.3\ 0/; A_{39L} = /0\ 0\ 0.7\ 0.3/;$$
$$A_{310L} = /0\ 0.3\ 0.7\ 0/; A_{311L} = /0.6\ 0.4\ 0\ 0/; A_{41L} = /0\ 0.3\ 0.7\ 0/; A_{42L} = /0\ 0.4\ 0.6\ 0/;$$
$$A_{51L} = /0\ 0.3\ 0.7\ 0/; A_{52L} = /0\ 0.6\ 0.4\ 0/; A_{53L} = /0\ 0\ 0.7\ 0.3/; A_{54L} = /0\ 0.3\ 0.7\ 0/;$$
$$A_{55L} = /0.7\ 0.3\ 0\ 0/.$$

6 The Algorithm for Obtaining the Degrees of Risk of the Analyzed Project

The algorithm for obtaining the degrees of risk of the analyzed project consists in assigning this project to one of the reference ones with assigning it the values of the degrees of risk. The following is the sequence of actions of this algorithm:

1. Selecting the first string of attribute values of the membership functions of the analyzed project from the database.
2. Calculation for this row of the sum of the squares of the differences between the values of the attribute membership functions of the analyzed project and the values of the elements of the etalon strings for all types of criteria.
3. Determination of the minimum value from the calculated amounts and the degree of risk of the etalon string corresponding to this minimum amount.
4. Selecting the second, third, and so on strings of attribute values of the membership functions of the analyzed project from the database.
5. Repeating operations pp. 2 and 3 algorithms until all strings of values of membership functions of attributes of the analyzed project are selected from the database and all values of risk degrees are obtained.

For example, for the analyzed project L from class A of project financing, the following values of risk degrees were obtained:

$R_{11L} = 1$; $R_{12L} = 2$; $R_{13L} = 3$; $R_{14L} = 2$; $R_{15L} = 1$; $R_{21L} = 2$; $R_{22L} = 1$; $R_{23L} = 1$; $R_{24L} = 1$; $R_{25L} = 2$; $R_{26L} = 2$; $R_{31L} = 2$; $R_{32L} = 2$; $R_{33L} = 3$; $R_{34L} = 3$; $R_{35L} = 3$; $R_{36L} = 1$; $R_{37L} = 3$; $R_{38L} = 2$; $R_{39L} = 3$; $R_{310L} = 3$; $R_{311L} = 1$; $R_{41L} = 3$; $R_{42L} = 3$; $R_{51L} = 3$; $R_{52L} = 2$; $R_{53L} = 3$; $R_{54L} = 3$; $R_{55L} = 1$.

7 Assessment of Financing Risk and Determination of the Rating of the Analyzed Project

The risk assessments of O_i financing for the credit requirements of the analyzed projects can be calculated using the following formula:

$$O_i = k_i \sum_{j=1}^{J} \frac{R_{ij}}{\max\{R_{ij}\}}, \tag{1}$$

where I is the number of the financing risk assessment; k is the significance coefficients of the i–th assessment set by experts, R_{ij} is the obtained values of the degrees of risk of the analyzed project in accordance with credit requirements.

The maximum financing estimate of the $\max\{O_i\}$ of the analyzed projects is proposed to be calculated using the following formula:

$$\max\{O_i\} = k_i * J, \quad (2)$$

where k_i are the significance coefficients of the i–th financing risk assessment, J is the number of credit claims within the i-th financing risk assessment.

The total assessment of the financing O_R of the analyzed project for all credit requirements:

$$O_R = \sum_{i=1}^{I} O_i, \quad (3)$$

When considering the financing rating of the analyzed project, it is proposed to use the ranking method:

1st rank—From 0 to (1/4) $\max\{O_R\}$ (Excellent borrower);

2nd rank—From (1/4) $\max\{O_R\}$ to (1/2) $\max\{O_R\}$ (Good borrower, but some minor comments);

3rd rank—From (1/2) $\max\{O_R\}$ to (3/4) $\max\{O_R\}$ (Satisfactory borrower and additional information required);

4th rank—From (3/4) $\max\{O_R\}$ to $\max\{O_R\}$ (Unsatisfactory borrower).

Using the example of the analyzed project L, let us consider in more detail the calculation of risk assessments for financing this project and assigning it an appropriate rating.

The maximum value of the risk degree max $\{R_{ij}\} = 5$ is set for the analyzed project L. Let's calculate the risk estimates of O_i financing for the values of risk degrees obtained in the previous paragraph. The values of the degrees of risk correspond to certain credit requirements.

For the first 5 values of the degrees of risk: $R_{11L} = 1$; $R_{12L} = 2$; $R_{13L} = 3$; $R_{14L} = 2$; $R_{15L} = 1$ the risk assessment of financing O_1 at $k_1 = 5$, $i = 1$, $J = 5$ in accordance with formula (1) is as follows: $O_1 = 5(1/5 + 2/5 + 3/5 + 2/5 + 1/5) = 9$. The maximum risk assessment of financing is calculated according to the formula (2): $\max\{O_1\} = 25$.

For the following 6 values of the degrees of risk: $R_{21L} = 2$; $R_{22L} = 1$; $R_{23L} = 1$; $R_{24L} = 1$; $R_{25L} = 2$; $R_{26L} = 2$ the risk assessment of O_2 financing at $k_2 = 4$, $i = 2$, $J = 6$ in accordance with formula (1) is as follows: $O_2 = 4(2/5 + 1/5 + 1/5 + 1/5 + 2/5 + 2/5) = 7.2$. The maximum risk assessment of financing is calculated according to the formula (2): max $\{O_2\} = 24$.

For the following 11 values of the degrees of risk: $R_{31L} = 2$; $R_{32L} = 2$; $R_{33L} = 3$; $R_{34L} = 3$; $R_{35L} = 3$; $R_{36L} = 1$; $R_{37L} = 3$; $R_{38L} = 2$; $R_{39L} = 3$; $R_{310L} = 3$; $R_{311L} = 1$ the risk assessment of O_3 financing at $k_3 = 1$, $i = 3$, $J = 11$ in accordance with formula (1) is as follows: $O_3 = 1(2/5 + 2/5 + 3/5 + 3/5 + 3/5 + 1/5 + 3/5 + 2/5 + 3/5 + 3/5 + 1/5) = 5.2$. The maximum risk assessment of financing is calculated according to the formula (2): max $\{O_3\} = 11$.

For the following 2 values of the degrees of risk: $R_{41L} = 3$; $R_{42L} = 3$ the risk assessment of O_4 financing at $k_4 = 3, i = 4, J = 2$ in accordance with formula (1) is as follows: $O_4 = 3(3/5 + 3/5) = 3.6$. The maximum risk assessment of financing is calculated according to the formula (2): $\max\{O_4\} = 6$.

For the following 5 values of the degrees of risk: $R_{51L} = 3$; $R_{52L} = 2$; $R_{53L} = 3$; $R_{54L} = 3$; $R_{55L} = 1$ the risk assessment of O_5 financing at $k_5 = 2, i = 5, J = 5$ in accordance with formula (1) is as follows: $O_5 = 2(3/5 + 2/5 + 3/5 + 3/5 + 1/5) = 4.8$. The maximum risk assessment of financing is calculated according to the formula (2): $\max\{O_5\} = 10$.

Next, using formula (3), we calculate the total financing estimate of the analyzed project L for all credit requirements: $O_R = 9 + 7.2 + 5.2 + 3.6 + 4.8 = 29.8$.

Also, using formula (3), you can calculate the total maximum financing estimate of the analyzed project L for all credit requirements: $\max\{O_R\} = 25 + 24 + 11 + 6 + 10 = 76$.

In accordance with the received estimates of financing risks for the analyzed project, we will conduct a ranking within the risk assessment range from 0 to 76. Let's introduce 4 ranks. The ranking results are shown in Table 1.

It follows from Table 1 that the analyzed project with a financing risk assessment of $O_R = 29.8$ is assigned a rank 2. This means that the analyzed project is good and the bank makes the appropriate decision (to finance the project or not). In case of non-financing of a project with a rank 2, it is possible to eliminate comments, thereby increasing the rating of the project and obtaining approval for financing.

8 Conclusion

According to this method, the possibility of non-repayment of the loan can be predicted, thereby assessing the credit risk. This method of logical–linguistic classification for assessing the risk of project financing as a class of specialized financing can be implemented in the form of a computer program that allows you to speed up the risk analysis of lending to the projects under consideration. At the same time, to assess the effectiveness and sustainability of project evaluations, computer simulation can be used, generating hundreds of possible combinations of project attributes and the values of their membership functions, taking into account their probabilistic distribution.

The method can also be used for other classes of specialized lending (object financing, commodity financing, financing of income-generating real estate, financing of real estate with unstable price parameters).

Acknowledgements The work was carried out within the framework of the state assignment of the Ministry of Science and Higher Education of the Russian Federation for the Institute of Problems in Mechanical Engineering of Russian Academy of Sciences.

References

1. Bulygina, O.V., Yemel'yanov, A.A., Yemel'yanova, N.Z., Kukushkin, A.A.: System analysis in management [Sistemnyy analiz v menedzhmente: uchebnik]. In: A.A. Yemel'yanov (ed.) Textbook, 2nd ed., p. 450. M.: FORUM: INFRA-M (2021). (In Russia)
2. Borisova, O.V., Malykh, N.Y.A., Oveshnikova, L.V.: Investments: Textbook and Practice for Universities [Investitsii: uchebnik i praktika dlya vuzov] 2nd ed., vol. 1. M.: Yurayt Publishing House (2024). (In Russia)
3. Boyarkin, G.N., Sheveleva, O.G., Tkachenko, A.L.: Selecting an investment object using analytical methods of risk assessment. Competitiveness Glob. World: Econ., Sci., Technol. 9–2(56), 23–27 (2017). (In Russia)
4. Vedmed' I.Y.U.: Analiz kolichestvennykh metodov otsenki riskov investitsionnykh proyektov. In: Probabilistic Mathematical Models of Economic Processes. XII Conference Russian Regions in the Focus of Change November 16–18, 2017, pp. 52–61. Yekaterinburg (2017). (In Russia)
5. Volkov, I.M., Gracheva, M.V.: Design Analysis [Proyektnyy analiz], p. 423. M.: Banks and exchanges, UNITY (1998). (In Russia)
6. Gareev, A.Z.: Risks during the implementation of investment projects. Innovative Sci. **10**(1), 30–33 (2016). (In Russia)
7. Gorodetskiy A. Ye., Tarasova I. L. (2010). Fuzzy Mathematical Modeling of Poorly Formalized Processes and Systems. [Nechetkoye matematicheskoye modelirovaniye plokho formalizuyemykh protsessov i system], p. 336. St. Petersburg: Publishing house of the Polytechnic University. (In Russia)
8. Yefremova, Y.A., Pryadkina, V.A.: Application of the Monte Carlo method for evaluating investment projects. Sci Community Students XXI Century. Econ. Sci.: Collect. Art. Based Mater. XXVII Stud. Intl. Sci. Pract. Conf. 12(27), 237–244 (2014). (In Russia)
9. Zhegalkin, I.I. (1928) Arithmetization of symbolic logic. Mathematical Collection [Arifmetizatsiya simvolicheskoy logiki Matematicheskiy sbornik]. 35(3–4), 335. (In Russia)
10. Zade, L.A. (1976). The Concept of a Linguistic Variable and Its Application to Approximate Decision Making [Ponyatiye lingvisticheskoy peremennoy i yego primeneniye k prinyatiyu priblizhennykh resheniy], p. 150. M.: MIR. (In Russia)
11. Korol'kova, Y.M. (2013). Risk Management: Project Risk Management [Risk-menedzhment: upravleniye proyektnymi riskami], p. 160. FGBOU VPO «TGTU», Tambov (In Russia)
12. Kulik, Yu.A., Volovich, V.N., Privalov, N.G., Kozlovsky, A.N.: Classification and qualitative assessment of risks of innovative projects. Notes Min. Inst. **197**, 124–125 (2012). (In Russia)
13. Mirkin, B.G.: Group Choice Problem [Problema gruppovogo vybora] p. 256. M.: Nauka, (1974). (In Russia)
14. Gorodetskiy, A.E., Tarasova, I.L., Krasavtseva, A.R.: Decision-making when assessing financing risks using the logical and linguistic method. In: Smart Electromechanical Systems. Mathematical and Software Engineering, Studies in Systems, Decision and Control, Springer Nature Switzerland AG, (2024). In this volume
15. Gorodetskiy, A.Y. (RU), Kurbanov, V.G. (RU), Tarasova, I.L. (RU). Method of Image Classification. Patent RU2756778C1 (2021)
16. Section I. Regulation of the Bank of Russia dated 08.06.2015 N 483-P (as amended on 04.15.2020) On the procedure for calculating the amount of credit risk based on internal ratings (together with Requirements for the quality of data used by banks to create and application of models for quantitative assessment of credit risk for the purposes of calculating capital adequacy standards).(In Russia).
17. Basel Committee on Banking Supervision. https://www.bis.org/baselframework/BaselFramework.pdf
18. Moskvin, V.A.: Risks of Investment Projects [Riski investitsionnykh proyektov], p. 256. M.: Kurs, INFRA-M, (In Russia) (2016)

19. Gorodetskiy, A.E., Tarasova, I.L.: Modeling of systems of logical-linguistic classification of images. Smart electromechanical systems. In: Smart Electromechanical Systems, Studies in Systems, , vol. 419, pp. 95–103. Book series Decision and Control (2022). https://doi.org/10.1007/978-3-030-97004-8_7

Formation of Control Actions

Forming an Emotional Response in a Robot

Viacheslav V. Kozlov

Abstract *Problem statement*: effective communication between robots and people requires taking into account the role of emotions in the mechanism of behavior control and joint practical activities. Recognizing human emotions and the robot demonstrating signs of an emotional response that are appropriate to the situation are important elements of successful interaction. Emotions have a wide range of manifestations, and trying to technically implement a robot to demonstrate signs of the entire spectrum of emotions is difficult. *Purpose of research*: analysis of ways to reduce the complexity of the task of demonstrating by a robot signs of an effective emotional response appropriate to the situation. In this case, a necessary condition should be the ability to unambiguously recognize the class of demonstrated emotions. Assessment of the possibility of choosing the type of emotional response based on the development of decision-making models with unclear initial data about the mixed emotional state of the person participating in communication. *Results*: The real requirements for the emotional manifestations of a robot during working interaction with a person, including within the framework of social robotics, are considered. Analysis of job descriptions that prescribe employee behavior when performing these jobs, expert opinions show that it is possible to significantly narrow the range of response emotions demonstrated by the robot, sufficient to complete the assigned tasks, reduce them to several clearly distinguishable classes, with technically accessible hardware implementation. Limiting the number of response emotions demonstrated by the robot allows the use of a simplified analysis of human emotions that generate the robot's response, the use of discrete models, and forming of data for decision-making using fuzzy logic. *Practical significance*: A mechanism for the formation of an emotional response of a robot participating in working interaction with a person is proposed. It allows you to limit and simplify the multidimensional meanings of feature describing emotions, and to use methods of logical–linguistic analysis to make decisions, reduces the likelihood of a robot's emotional reaction not matching the situation.

V. V. Kozlov (✉)
Institute for Problems in Mechanical Engineering of the Russian Academy of Sciences, Saint-Petersburg, Russia
e-mail: vvk26@mail.ru

I. L. Tarasova and B. A. Kulik (eds.), *Smart Electromechanical Systems*, Studies in Systems, Decision and Control 544, https://doi.org/10.1007/978-3-031-64277-7_17

Keywords Interaction · Social robotics · Informative emotions · Fuzzy logic · Emotional response

1 Introduction

Progress in the field of artificial intelligence, supported by successes in the development of electromechanical components and various sensors, has expanded the scope of robots to areas of activity directly related to both human service—social robotics, and to working in a team with a person..Integration of robots into the scientific, technical and social sphere of society has become a specific task of our time. The development of principles for designing the appearance and behavior of robots, taking into account the functional tasks facing it and the characteristics of their clients in the case of social robotics, or a human partner in joint work, comes to the fore. Work is being intensively carried out to study the problem of the influence of the humanoid appearance and behavior of an android on the nature of interaction with a person. Study methods and mixed methods for a comprehensive assessment of the interaction between a robot and a person are being developed [1].

At the same time, numerous examples show that it is necessary not only to develop and form the specific properties of a robot, but also to prepare human psychology for communication with a robot. In the case of positioning a robot as an equal partner, the key factor is the human perception of it. For his part, a person must adjust and adapt to the robot interacting with him. This need is evident from well-known cases of attempts to establish equal relationships, for example with the HitchBOT robot, which was created as a cute hitchhiker who is able to carry on a simple conversation and record the journey in photos. HitchBOT's journey after Canada, Holland and Germany ended in Philadelphia. An unknown person smashed the robot without any reason and with particular cruelty [2]. To solve this problem, the Society for the Prevention of Cruelty to Robots (ASPCR) was created in the United States.

One of the objectives of the Society is to discuss the moral and ethical consequences of the appearance of intelligent artificial beings in our world and the responsibility associated with their creation [3]. The socialization of robots is proposed based on a model in which behavior between people is transferred to robots. The situation is complicated by the perception of the robot as a device that can be ignored. Interest in communicating with a robot can fade especially quickly if there is a belief that it does not understand the emotional state of a person. Removing barriers to communication requires robots to demonstrate the initial elements of a human-like psyche; they must adapt to the behavior and mental characteristics of the particular person with whom they interact. In order to carry out high-quality interaction, the robot must understand who is in front of it, as well as assess the emotional state of the person. According to the recognized emotion, the robot will be able to choose the most effective communication strategy and will be able to demonstrate the required emotional response.

2 Assessing Emotions

The importance of the problem is emphasized by the development of affective computing technologies that enable interaction between a person and a computer using emotions [4]. Affective computing is based on data from scientific disciplines such as computer science, psychology and cognitive science. Affective computing has received significant attention, with efforts aimed at empowering robots to exhibit appropriate emotional behavior in order to enhance natural interactions with humans. Simulating emotional traits increases the effectiveness and adaptability of collaboration in collaboration, and the ability to demonstrate recognizable emotional expressions in social robotics is essential for believable communication. However, the basis for effective emotional behavior of a robot is the ability to correctly determine and interpret the emotional state of the person with whom it interacts. The problem is that the complexity of human emotions does not allow us to unambiguously fully describe their meaning and the functions they perform.

Currently, numerous works are aimed at developing algorithms for classifying emotional states according to various characteristics, such as facial expression, body language, and voice. An obvious way to assess emotions is to analyze facial expressions. Methods are proposed based on the study of the geometric characteristics of facial landmarks and the analysis of information about the displacement of landmarks in time. To increase reliability, thermal imaging cameras, distance sensors, and analysis of three-dimensional information are used. Body posture, movement, and gestures are natural ways to infer a person's emotional state. The voice is also an informative channel about the emotions of our interlocutor. In this case, both acoustic characteristics are studied—pitch, energy, spectral components, and semantic content. The choice of informative features and their correlation is critical. A significant number of methods have been proposed for processing the results obtained to solve the problem of determining the observed emotional state, including using neural networks. The main difficulty of recognition is the presence of complex variations within emotions, the overlap of different emotions in observed behavior [5, 6]. There are two main models for representing emotions. This is a model where emotions are represented by a small number of discrete classes, and multidimensional models in which emotions are defined by continuous values of the features that describe them. Multidimensional models of emotions consider all emotions in a multidimensional space, a limited number of variables (coordinates), for example: negativity—positivity, strength—weakness, activity—passivity. Another model offers a variety of other three dimensions (sensations): pleasure-unpleasure, relaxation-tension, and calm-excitement.

From the point of view of the recognition process, the ability to identify individual discrete emotions is ideally the simplest method. However, as the number of emotions considered increases, it becomes increasingly difficult to distinguish between their classes. Simplification of the representation leads to an increase in errors in their definition. For example, using the theory of "basic emotions". It assumes that any person is able to recognize and use several basic emotions: joy, sadness, anger, fear and disgust. All of these conditions are closely related to human behavioral characteristics and can be identified. However, a more detailed analysis shows that emotions cannot be uniquely defined and their expression is determined by the cultural and individual characteristics of a particular person. For example, a smile is easily recognized by algorithms and identified as a state of happiness. However, in reality, the function of a smile can change depending on a person's social environment and carry a different emotional mood. Similar problems arise when interpreting gestures and "body language." There is no confidence in their uniqueness and universality.

3 Level of Expression of Emotions

All of the above demonstrates both the importance of emotions and the existing limitations of their use for interaction in a human–robot system. If the task is to demonstrate and recognize the entire spectrum of emotions, then it is obvious that at the moment it is impossible both to reliably recognize emotions and to technically demonstrate emotions by a robot. However, the question arises: why should a robot demonstrate the entire range of emotions available to a human? The robot does not have universal human problems associated with the struggle for existence, caring for loved ones, procreation, and ensuring self-development. The most complex emotional manifestations ensure the evolutionary development of a person, which is inaccessible to a robot, since the robot does not have a natural source of generating changes in the process of freely exploring the surrounding world.

Analyzing the real areas of activity proposed for a robot, one can see that to effectively perform work that requires communication with a person, it is quite enough limited number of rough emotion models required for the robot to demonstrate. Just look at the advice and job descriptions of workers in various fields of activity. For example, consider the requirements for the behavior of a doctor [7]. The behavior of a doctor, especially when he must act quickly and decisively, should be rather indifferent and cold. Sympathy should be laconic. There should be a display of moderate empathy. The basis for expressing empathy is pauses, support, but also refusal of premature encouragement. When asking a question, the doctor must listen carefully and calmly to the patient, always wait for an answer, and express his thoughts clearly, simply, and intelligibly. Manifestations of arrogance, neglect or humiliating treatment of the patient are excluded. Firm calm, attentiveness, and optimism are required. Let's look at the requirements for a bartender in the same way [8]. He is required to constantly maintain neutrality and display an energetic positive mood. Patience and calm respect for the client. If there is aggression on the part of the client, it is necessary to reduce activity, speak more quietly, calmly, and more slowly than usual. If aggression intensifies, you need to ask for help, but do not get involved in the conflict, showing neutrality. If we consider the requirements for behavior that must be demonstrated for successful joint production work, then employees are required to show calm, be able to listen, and show empathy [9].

We can continue to analyze the requirements for the behavior of workers in various fields of activity, but still, everywhere they require the manifestation of a very limited range of reciprocal positive emotions, regardless of the situation. It can be seen that from a good employee to successfully perform his functions, only a few simple and easily distinguishable patterns of manifestation of neutral or positive emotions are required, complex emotional experiences of high intensity are excluded. These requirements can be transferred to the robot, as a good worker, by offering it a limited number of models of emotional response. A limited set of response emotion models makes it possible to simplify the task of determining the emotional state of a person interacting with a robot. For example, it is generally accepted that the most common emotions that a person may experience during intense work are anger, irritation, dissatisfaction, anxiety, and hostility. Since the robot's response emotions when manifesting these complex states are selected from a limited number of models, the robot's recognition of these considered complex human emotions can be reduced to a classification of almost one generalized type.

4 Modeling the Necessary Emotions

When creating a system for modeling the response emotions of robots to the behavior of a person interacting with him, the most important task is making a decision based on the analysis of incomplete and insufficiently reliable information [10]. Most of the existing methods for modeling phenomena are effective only if comprehensive information about these phenomena is available. If such conditions are not met, then the presence of only approximate, inaccurate information cannot provide an unambiguous characteristic of the phenomenon under study and lead to the construction of an unambiguous model. In this case, the result of the study is a decision determined by enumerating the values of those parameters about which information was incomplete. This leads to the creation of fuzzy models based on the apparatus of the theory of fuzzy sets and fuzzy logic, and the main efforts are aimed at simplifying them and using a limited number of not all available parameters, but only some of the most important and necessary for specific conditions.

One of the decision-making options may be a fairly fully studied mechanism for selecting optimal solutions in conditions of incomplete certainty of the logical–linguistic type. For this purpose, a fuzzy inference algorithm is used. It includes determining, based on given values of factors, their degree of belonging to various terms of the corresponding linguistic variables, carrying out fuzzification of the initial values of factors. Next, using the knowledge base and performing operations on fuzzy sets, it is possible to construct a fuzzy set of output indicator values, provided that the input factors have specified values. At the final stage, using the resulting fuzzy set, it is possible to defuzzify it and obtain a specific value.

Let's consider an example of the emotional response of a robot bartender. The example is tentative due to the lack of necessary knowledge bases. These databases must be created by experts—specialists in the field in question. They include both an assessment of the client's possible behavior based on signs of expression, and the level of service corresponding to the behavior, as well as the emotions and actions of the robot bartender necessary for this level of service. The required level of service is set by the emotional behavior of the client and determines the emotional response of the robot. The emotional state that determines the client's motives for behavior is determined through expression channels—facial expressions, gestures, speech. Currently, significant progress has been made in this area. Recognition of basic emotions by facial expression reaches more than 80% on a set of classifiers. Based on gestures and body position, recognition of basic emotions also reaches more than 80%. The emotions of our interlocutor based on the semantic content of what he is saying, as well as analysis based on 32 acoustic signs, basic emotions were determined with a probability of up to 80%. For recognition, methods of preliminary information processing, deep learning, and neural network functions are used. These results were achieved in laboratory conditions, in the absence of interference and other interfering factors. However, these emotion recognition systems only used one expression channel; multi-channel systems should be more successful [11].

Let's introduce the variable "required level of customer service" by the robot bartender, with values from "ask for help from security" to "maximum effective service." To determine the required level of service, we introduce for the client the characteristic variables "level of aggression", that is, aggressive—good-natured, and "level of intrusiveness", that is, intrusive—indifferent. We believe that these variables are sufficient to describe the choice of service level. Recognition of these types of emotional behavior due to their small number, the absence of the need for careful differentiation, as when compared with the case of recognition of basic emotions, is simpler and more reliable. Using several expression channels, having visual and acoustic standards of the level of the entered variable, based on the analysis of expression features, it is possible to fuzzify the initial values. To construct a fuzzy set of output indicator values, a knowledge base is used that connects input factors with given values. The database is created on the basis of expert assessments, and provides estimated values of the influence of the combination of factors "level of aggression" and "level of intrusiveness" on the required level of service. Using the resulting fuzzy set, it is possible to defuzzify it and obtain a specific assessment on the scale of the required level of service. The choice of a robot's emotional reaction can be represented as solving an inverse problem. It is known that in order to perform his functions well, a bartender needs to show a very limited amount of emotion. This allows us to introduce the variable "friendliness" to describe the emotional behavior of the robot bartender, that is, from calm rejection, dislike, to calm friendliness. To express the signs of a robot's expression—facial expressions and speech—a minimum set of their values is sufficient. Simple expression standards make it possible to achieve technical reproducibility of the required range of facial expressions and speech features of a robot. Using a previously created knowledge base that determines the compliance of the technical reproduction of expressions corresponding to a certain required level of customer service, the robot will be able to demonstrate the required emotional response. Thus, knowing from the client's behavior the required level of his service, using this value we find the necessary signs of emotional expression for the robot bartender. However, since the robot is an actuator, and technical failures in its operation are possible, it is desirable to introduce feedback and control the robot's performance of specified expressions of emotions. This complicates the algorithm, but allows for increased reliability.

The diagram of an emotional robot may look as shown in Fig. 1. The channel for determining the client's state includes sensors 1 of different channels for assessing expression, blocks 2 process sensor signals and transmit them to fuzzification blocks associated with the knowledge base 5 of client behavior assessments. After fuzzification, the information enters the fuzzy logical inference block 4 and then to the defuzzifier 6, which, based on information from the knowledge base 5, makes an assessment of the required level of customer service. Block 7 allows this score to be shifted, which simulates some of the robot bartender's personality. For example, if you shift the assessment of the client's behavior towards a higher level, the robot's behavior will be more liberal. Or, keeping in mind that many bars have regular visitors, implement the robot's memory for customers and enter an appropriate preference for the level of service.

Block 8 generates the necessary control signals for technical devices 9 depicting types of expression. To do this, he uses the knowledge base 16 obtained by assessing the quality of emotions displayed by the robot's technical means. This knowledge base can be filled using the channel for monitoring the robot's display of emotions, and it is used to initially set the robot's emotional state. The channel for monitoring the emotions shown by the robot works similarly to the channel for monitoring the client's state. Signals from sensors 15 different channels for determining the emotions shown by the robot are sent to the primary signal processing blocks 14, then to the fuzzification blocks 13 associated with the knowledge base 12 of assessments of

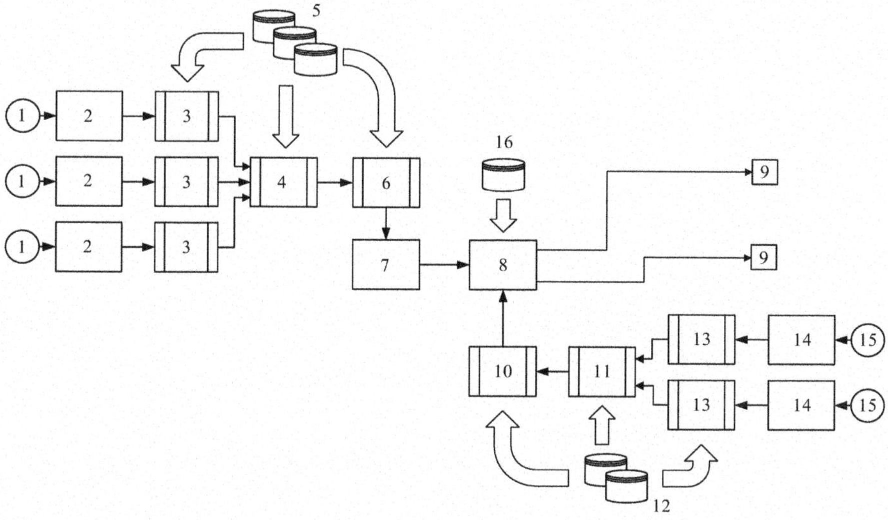

Fig. 1 Example of a functional diagram of an emotional robot

the robot's manifestations of emotions. After fuzzification, the information goes to the fuzzy inference block 11 and then to the defuzzifier 10, which, based on the information in the knowledge base 12, makes an assessment of the emotions shown by the robot, calculating what level of customer service they correspond. In block 8, a comparison is made between the calculated required level of service and the level of service corresponding to the actual emotions shown by the robot. If necessary, the expression of emotions is corrected. Taking into account the specifics of a bartender's work, this model can be expanded. It is possible to introduce a delay in the manifestation of an emotional response, expecting a change in the client's behavior after the necessary reaction of the robot.

4.1 Conclusions

Analysis of human emotions, their positioning in order to model in a robots meet significant difficulties. However, if we consider the combination of emotional manifestations that is really necessary for the effective joint interaction of a person and the robot, measured by well -defined signs, the degree of complexity of the task can be reduced. For this, it is more important not to classify emotions according to the chosen model, but to determine the influence of a person's emotional state on his possible behavior. The reciprocation by the robot of the minimum sufficient set of emotions can be determined on the basis of known mathematical methods for processing fuzzy information.

Acknowledgements The work was carried out within the framework of the state assignment of the Ministry of Science and Higher Education of the Russian Federation for the Institute of Problems in Mechanical Engineering of Russian Academy of Sciences.

References

1. https://rg.ru/2020/08/05>v-yaponii-mashiny
2. https://edition.cnn.com/2015/08/03/us/hitchbot-robot-beheaded-philadelphia-feat/index.html
3. http://www.aspcr.com
4. Picard, R.: Affective Computing. MIT Press, Cambridge, MA (1997)
5. James, W.: Psychology-M.: Pedagogy (1991)
6. Ilyin, E.P.: Emotions and Feelings. S-Pb.:Piter (2001)
7. https://orgzdrav.vshouz.ru/ru/jarticles_orgzdrav/11.html?SSr=050134d7bd15ffffffff27c__07e803110f051e-1e3d
8. https://ampravda.ru/2015/06/16/058134.html
9. https://brainapps.ru/blog/2023/11/kak-upravlyat-emotsiyami-komandyia-sekretyi/

10. Gorodetskiy, A.E., Tarasova, I.L.: Nechetkoye matematicheskoye modelirovaniye plokho formalizuyemykh protsessov i system [Fuzzy Mathematical Modeling of Poorly Formalized Processes and Systems] p. 335. St. Petersburg, Polytechnic, University Publication (2010). (In Russian)
11. https://www.frontiersin.org/articles/10.3389/frobt.2020.532279/full

Autocollimation System with Extended Distance for Angular Rotation Measurements in SEMS Applications

Mikhail M. Nikitin and Igor A. Konyakhin

Abstract *Problem statement*: When working with large-scale objects a group of robots is efficient to use. Robots may be combined into SEMS for synchronous operations like welding, cutting or moving objects. Such operations imply restrictions on the tool's angular positioning precision. To measure the angular rotation of controlled elements with high accuracy, autocollimation systems can be applied. The systems currently available on the market, do not possess long working distance, therefore, the problem is relevant. *Purpose of research*: This study focuses on the special type of an autocollimation system reflector: mirror tetrahedron with one of flat surfaces replaced with the cylindrical ones. This type of reflector has a number of advantages over other trihedral elements with a flat angle between two surfaces deviating from 90°; or flat mirrors; or a right-angle prism, which provides an extended measurement range and working distance. *Results*: Mathematical study shows that the autocollimation system proposed has the measurement range reaching ± 20°, which is more than ten times increase in comparison with its analogues—Taylor Hobson Ultra Autocollimator: ± 30′, Davidson Optronics D-638 Autocollimator: ± 20′, Opto Sigma TM-2070: ± 70′). The working distance can be extended to 30m. *Practical significance*: The autocollimation system described in the paper can be applied for a group of robots, operating on a large-scale object. For instance, it can be welding or cutting tool positioning. These tools are placed on the moving parts of SEMS, and it is critical for the initial position of the tool to be determined with high accuracy. The autocollimation system can ensure high precision for angular positioning of the SEMS moving elements, which are placed at different distances from the base.

Keywords Smart electromechanical systems · Autocollimation system · Roll, pitch and yaw measurement · Computer simulation · Cube-corner reflector · Cylindrical mirror

M. M. Nikitin (✉) · I. A. Konyakhin
ITMO University, Saint-Petersburg 197101, Russia
e-mail: mmnikitin@itmo.ru

I. A. Konyakhin
e-mail: iakonyakhin@itmo.ru

I. L. Tarasova and B. A. Kulik (eds.), *Smart Electromechanical Systems*, Studies in Systems, Decision and Control 544, https://doi.org/10.1007/978-3-031-64277-7_18

1 Introduction

The SEMS technology rise has provoked the need in precise working tool or working robot parts position measurement. Autocollimation systems can be used to achieve such type of precision [1–3]. Let us take a look at the example of autocollimation method application in the case of a group of robots, performing welding operation on a large-scale object. Figure 1 demonstrates the described application. The robots have manipulators, which control the working tool position. Special reflectors (2) are placed near each working tool of a robot. The autocollimator (1) is placed on the static base. Initial positioning is crucial for this type of operation. That is why, to calibrate each robots working tool angular position, the following steps are required. First, the robot moves its working tool at the autocollimator field of view. Afterwards, the autocollimator generates a beam of light, which is reflected back by a reflector. The angles measured from reflected beam are transferred to the robot to correct its working tool position. When all robots are calibrated, they can start the operation.

This example shows that for coordinated work robots need the autocollimation system with long working distance. Existing solutions [4, 5]—flat mirror, right-angle prism or mirror pyramid—have the following problem: information about the angle of a controlled object is acquired from reflected beam coordinates. This fact justifies that if the system working distance is enlarged, its measurement range becomes smaller. To solve this problem, a trihedral reflector with cylindrical surface can be applied [6]. This type of reflector has unique properties. The image formed on the autocollimator sensor is cross-shaped, and the information about the angles of controlled object rotation is presented in the rotation of lines in the image. The reflector described in [6] is challenging to use for three-dimensional angular measurement, since there are only two lines on the image, therefore, we propose to use a reflector with the configuration of cylindric surfaces [7, 8] presented in Fig. 2.

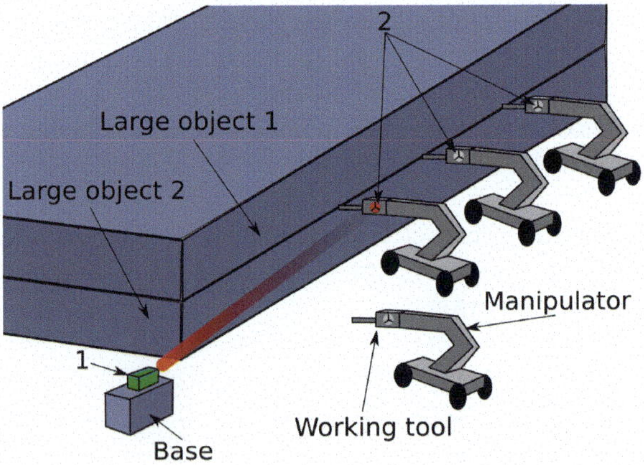

Fig. 1 Example of welding operation done by group of robots controlled with autocollimation method

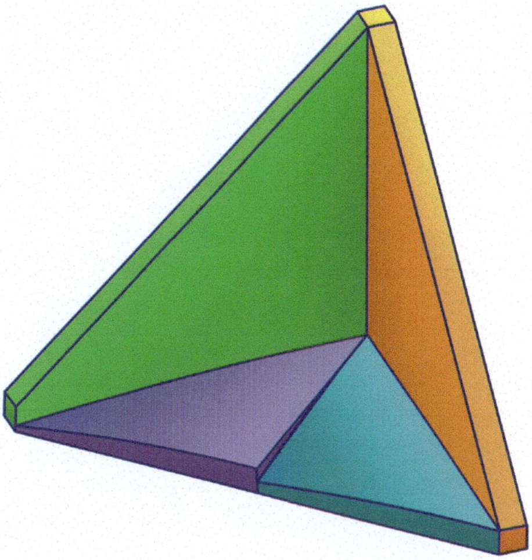

Fig. 2 Trihedral reflector with cylindrical surface configuration

2 Mathematical Model

Let us describe the process of mathematical modeling for this reflector. Reflection matrices of flat surfaces are (1):

$$Ry = \begin{pmatrix} 1 & 0 & 0 \\ 0 & -1 & 0 \\ 0 & 0 & 1 \end{pmatrix}; Rx = \begin{pmatrix} -1 & 0 & 0 \\ 0 & 1 & 0 \\ 0 & 0 & 1 \end{pmatrix}. \quad (1)$$

For cylindrical surfaces the reflection matrix is (2):

$$Rzc = \begin{pmatrix} 1-2N_x^2 & -2N_xN_y & -2N_xN_z \\ -2N_xN_y & 1-2N_y^2 & -2N_yN_z \\ -2N_xN_z & -2N_yN_z & 1-2N_z^2 \end{pmatrix}, \quad (2)$$

where N_x, N_yN_z can be found from (3)

$$N_x = \cos(\alpha)\sin(\delta), N_y = \sin(\alpha)\sin(\delta); N_z = \cos(\delta). \quad (3)$$

In the case of two cylindrical surfaces, we apply the following parameters: $\alpha_1 = -45°$, $\alpha_2 = 0°$ и $\delta_{max} = 10°$ (Fig. 3).

When describing the reflected beam, the sequence of reflections is critical. For the specified parameters, and accounting the fact that matrices for 1st and 2nd flat surfaces are commutative, in the image there should be eight different parts. However, the physical properties of cylindrical surface *32* differ from the properties of *31* so that reflection sequences *2–32-1* & *32–2–1* & *32–1-2* with *1–32–2* & *1–2–32* &

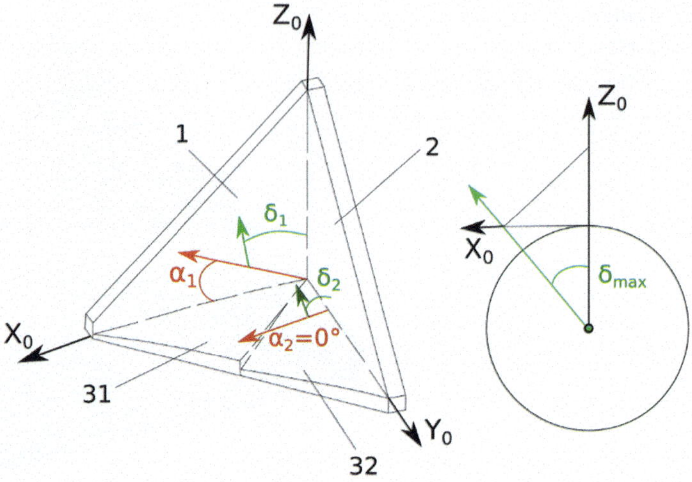

Fig. 3 Visualization of mathematical model of the reflector

2–1–32 result in only one line, determining that there are only six permutations. Let us look at the image in Fig. 4. The reflection sequences are distributed the following way: red line: *31–2–1*, orange: *1–2–31*, green: *1–31–2*, blue: *2–31–1*, violet: *2–32–1*, cyan:*1–32–2*.

To summarize, for each sequence of reflections, reflection matrix can be described as (4):

$$R_{abc} = Ra \cdot Rb \cdot Rc, \tag{4}$$

where—Ra, *Rb* и *Rc*—are one of the matrices (1)–(2).

For a better representation of the coordinate system and its coherence with the coordinate system of the image, we can use translation matrix (M_p) described by (5):

Fig. 4 Reflection from a reflector

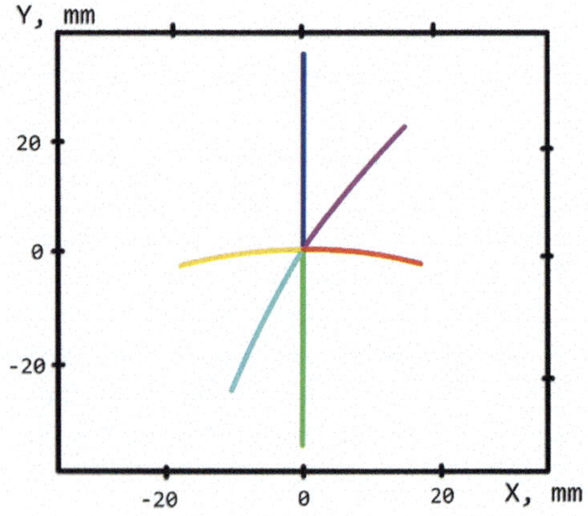

$$M_p = \begin{pmatrix} \frac{-1}{\sqrt{2}} & \frac{1}{\sqrt{2}} & 0 \\ \frac{-1}{\sqrt{6}} & \frac{-1}{\sqrt{6}} & \frac{2}{\sqrt{6}} \\ \frac{1}{\sqrt{3}} & \frac{1}{\sqrt{3}} & \frac{1}{\sqrt{3}} \end{pmatrix}. \tag{5}$$

The matrices of rotation with Eulers angles are taken from equations in [7]. Finally, the matrix of reflected beam for any sequence of reflections *a–b–c* and for the set of Eulers angles look like (6):

$$B_{abc\Theta} = \mathrm{M}_\Theta \cdot M_p \cdot R_{abc} \cdot M_p^T \cdot M_\Theta^T \cdot A, \tag{6}$$

where—M_Θ—rotation matrix as it is in [7], A—vector of falling beam normal to the front surface $A = (0, 0, -1)$.

Coordinates on the image can be found in (7)–(8):

$$Y = f \cdot \frac{B_{abc\Theta}[2]}{B_{abc\Theta}[3]} \tag{7}$$

$$X = f \cdot \frac{B_{abc\Theta}[1]}{B_{abc\Theta}[3]}, \tag{8}$$

where—f—focal length of autocollimator, $B_{abc\Theta}[1]$, $B_{abc\Theta}[2]$, $B_{abc\Theta}[3]$—projections of vector $B_{abc\Theta}$ to X (1), Y (2), Z (3) axes.

3 Image Changes

Let us take a look at different image changes due to different angles of reflectors rotation in the cases of the X_1, Y_1, Z_1 axes (Fig. 5).

In the case of rotation along the Z_1 axis, the whole image rotates to the same angle equal to Θ_3 (Fig. 6).

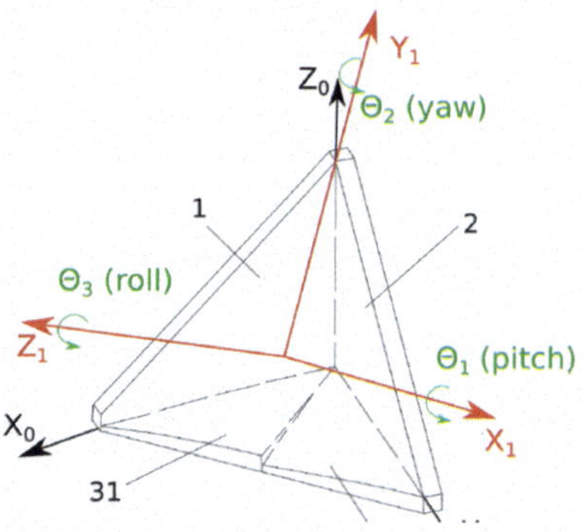

Fig. 5 Two coordinate systems

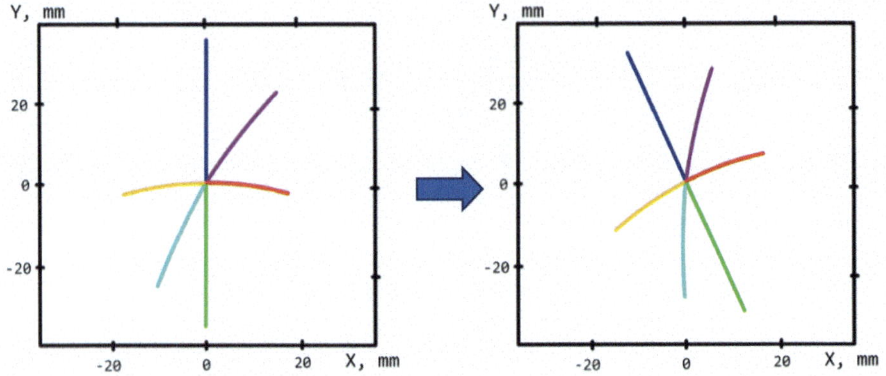

Fig. 6 Rotation to the angle $\Theta_3 = 20°$

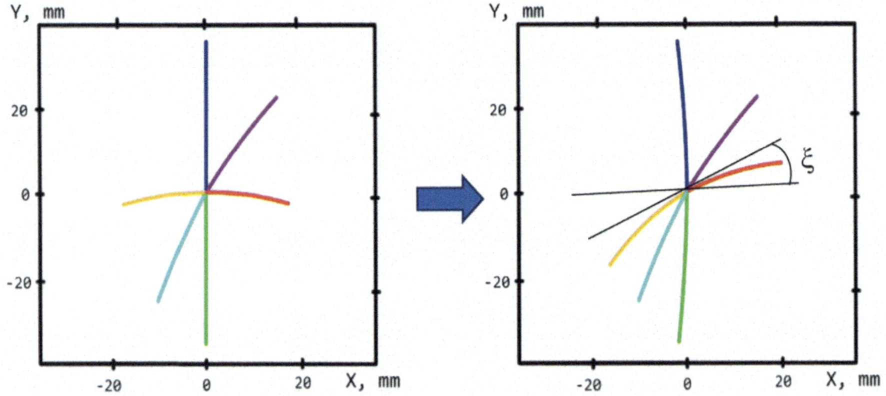

Fig. 7 Rotation to the angle $\Theta_2 = 20°$

This image change is due to the fact, that the Z_1 axis coincides with autocollimators optical axis.

In the case of Θ_2 (Fig. 7), the part of image, which was initially horizontal, rotates to angle $\xi \approx \sqrt{2}\Theta_2$.

When turning to Θ_1 (Fig. 8), the angle of the third line changes from its initial position ζ_0.

4 Measurement Algorithm

All three angles can be determined based on the following steps (Fig. 9):

(1) For the initially vertical line, angle Θ_3 is calculated (relative to Y).
(2) Angle Ψ should be measured to find Θ_2.

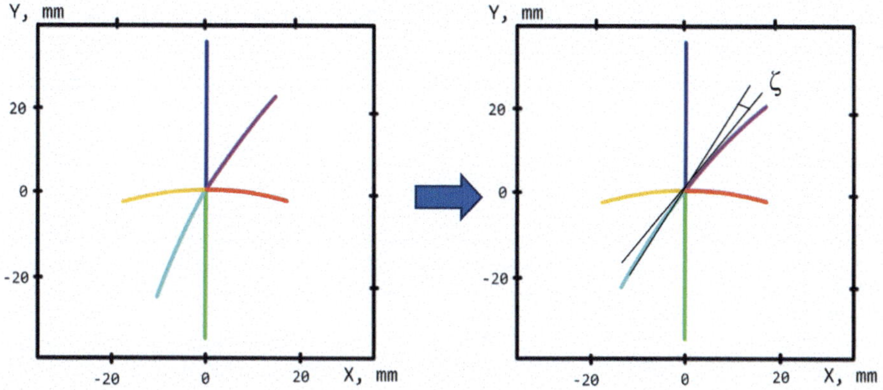

Fig. 8 Rotation to the angle $\Theta_1 = 20°$

Fig. 9 Algorithm for measuring angles on the image

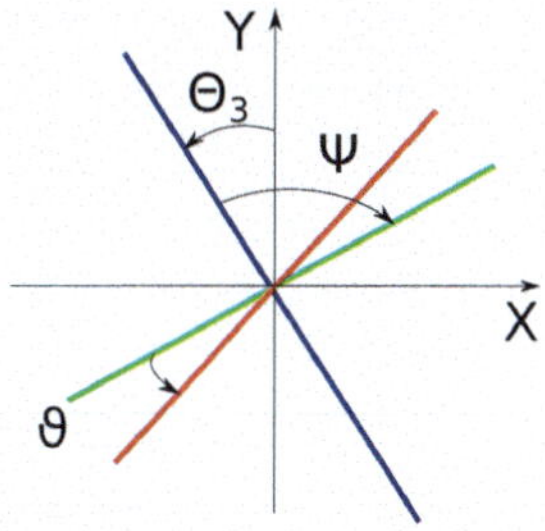

$$\Psi = \frac{\pi}{2} - arctg\left(\sqrt{2}\sin\Theta_2\right)$$

(3) Angle ϑ should be measured to find Θ_1.

$$\vartheta = \frac{\pi}{3} + arctg\left(\frac{\sqrt{6}\sin\Theta_1}{2}\right)$$

To calculate angles of lines in the image, they can be approximated with straight lines, and the method described in [8] can be used.

5 Conclusion

This paper describes the autocollimation system featuring a trihedral reflector with cylindric surface configuration. The main benefit for the system is enlarged working distance. Mathematical modeling shows that precision of 11″ can be achieved to solve the problem. This system can be applied to control the initial position of a working tool for the group of robots, which perform welding or cutting operations on large-scale objects.

References

1. Nogin, A.A., Konyakhin, I.A.: The position monitoring robotic platforms of the radiotelescope elements on base of autocollimation sensors. Stud. Syst., Decis. Control **174**, 297–304 (2019)
2. Nogin, A.A., Konyakhin, I.A.: Optoelectronic SEMS for preventing object destruction. Stud. Syst., Decis. Control **261**, 199–204 (2020)
3. Konyakhin, I.A., Dang, D., Turgalieva, T.V.: Autocollimation sensor for high-accuracy line and angular localization of SEMS blocks in technogenic environment. Stud. Syst., Decis. Control **419**, 221–233 (2022)
4. Yanhe, Y., Sheng, C., Yanfeng, Q.: Design, fabrication, and verification of a three dimensional autocollimator. Appl. Opt. **55**(35), 9986–9991 (2016)
5. Turgalieva, T., Konyakhin, I.: Research of autocollimating angular deformation measurement system for large-size objects control. Proc. SPIE **8788**, 878832 (2013)
6. Konyakhin, I.A.: Trihedral reflectors for three-axis angular autocollimation measurements. In: Konyakhin, I.A., Renpu, L., Smekhov, A., Kleshchenok, M. (eds.) In Latin America optics and photonics conference, OSA technical digest (online) (Optical Society of America) (2014). Paper LTh4A.49, 0.187/0.09
7. Nikitin, M., Konyakhin, I., Li, R., Xiao, H.: Benefits of trihedral reflectors with two cylindrical surfaces for three-dimensional autocollimation measurements. Proc. SPIE **12314**, 123140P (2022)
8. Nikitin, M., Konyakhin, I., Renpu, L., Xiao, H.: Angular determination algorithm for three-dimensional autocollimation measurements. Proc. SPIE **12769**, 1276913 (2023)

Polarization Imaging of Pathological Biological Tissues Using a Robotic System Based on Multi-element Sensors

Ruslan D. Khlynov, Victoria A. Ryzhova, Ekaterina A. Tsygankova, Artemii Y. Zhdanov, Igor A. Konyakhin, Valery V. Korotaev, Sergey N. Yarishev, Todor S. Djamiykov, and Marin B. Marinov

Abstract *Problem statement*: The spread of cancer is one of the most serious problems in medicine, therefore, in order to combat this disease, it is critically important to detect it at an early stage of development. However, the development of effective diagnostic methods for detecting pathological biological tissues and monitoring their development is a difficult task. Currently, optoelectronic methods are actively developing and improving, which make it possible to obtain high-quality, accurate and objective data on the state of biological tissues. This significantly increases the reliability of the diagnostic process and allows you to more accurately determine the degree of malignancy of the pathology. Therefore, in order to solve this problem, it seems relevant to propose and investigate optoelectronic methods for diagnostics, on the basis of which robotic systems based on SEMS modules can be created. *Purpose of research*: Quantitative analysis of the degree of malignancy of pathology using

R. D. Khlynov (✉) · V. A. Ryzhova · E. A. Tsygankova · A. Y. Zhdanov · I. A. Konyakhin · V. V. Korotaev · S. N. Yarishev
ITMO University, Saint-Petersburg 197101, Russia
e-mail: ruslan384ruslan3@gmail.com

V. A. Ryzhova
e-mail: victoria_ryz@itmo.ru

A. Y. Zhdanov
e-mail: artemichzdanov@gmail.com

I. A. Konyakhin
e-mail: iakoniakhin@itmo.ru

V. V. Korotaev
e-mail: vvkorotaev@itmo.ru

S. N. Yarishev
e-mail: sniaryshev@itmo.ru

T. S. Djamiykov · M. B. Marinov
Technical University of Sofia, Sofia 1756, Bulgaria
e-mail: tsd@tu-sofia.bg

M. B. Marinov
e-mail: mbm@tu-sofia.bg

I. L. Tarasova and B. A. Kulik (eds.), *Smart Electromechanical Systems*, Studies in Systems, Decision and Control 544, https://doi.org/10.1007/978-3-031-64277-7_19

the polarization-optical method, which is based on the registration of changes in the structure of biological samples with anisotropic properties. *Results*: The clinical effectiveness of the statistical multivariate analysis of the coordinate distributions of the invariants of the Mueller matrix obtained on the basis of the formation of digital images by laser imaging polarimetry in relation to the problem of differentiation of pathological conditions of biological tissue has been revealed. Azimuthally stable invariants of the Mueller matrices for samples of biological tissue of different organs and the degree of malignancy have been determined. It is proved that they characterize the polarization manifestations of porphines against the background of linear birefringence and optical activity of proteins of biological tissues. A comparative study of the diagnostic capabilities of the Mueller matrix polarimetry method of biopsy of gastric adenocarcinoma and lung metastasis with varying degrees of malignancy of the neoplasm was carried out. *Practical significance*: The creation of robotic systems based on SEMS modules will automate the process of diagnostics and analysis of optical properties of biological tissues. This will ensure automation and acceleration of the diagnostic process, as well as increase the efficiency of the medical staff.

Keywords Biological tissue · Degree of differentiation · Diagnosis · Pathology · Imaging polarimetry · Mueller matrix · Statistical analysis · SEMS · Robotic system

1 Introduction

One of the main priorities of modern medicine is the early diagnosis of cancer, which is characterized by uncontrolled and aggressive development of malignant cells in various organs and tissues of the body.

Effective diagnosis of cancer is a difficult task, because a tumor begins with a mutation of one cell, remains localized at an early stage of its development and is detected only after reaching a certain size. At this stage, the probability of cancer spreading through the body (metastases) increases rapidly, and at the same time, treatment becomes more difficult [1, 2].

Almost all human organs consist of glandular and epithelial cells. The oncological process in them that leads to the development of malignant neoplasm is called adenocarcinoma and can appear in any internal organ of the human body [3].

Objective determination of the location of the carcinoma focus and the further direction of metastasis to other organs is one of the directions of the development of biomedical sciences.

In particular, in order to choose the optimal treatment tactics, it is necessary to determine the degree of malignancy of the neoplasm, which is quantified by the degree of differentiation of the biological tissue and which is usually designated by the letter “G” (grade) [4].

Measuring the degree of differentiation and, based on it, monitoring and analyzing the evolution of tissue throughout the oncogenesis process are important to reduce the threshold for cancer detection. At the same time, it is important not only to develop new technologies, but also to ensure their availability in clinical practice.

Currently, there is an active development and improvement of optoelectronic methods that allow us to obtain high-quality and objective data on the state of biological tissues and which can be used in the diagnosis of cancer. A promising direction is the use of complex robotic systems based on SEMS modules (Smart Electromechanical Systems), which integrate various optoelectronic technologies and ensure high accuracy and reliability of diagnostic procedures thanks to special algorithms and software [5–8].

They can also be equipped with sensors that allow you to monitor the patient's condition in real time, which significantly increases the safety and effectiveness of diagnosis.

Currently, polarization-optical methods are widely used in the study of the structural features of biological tissues. The polarization segregation of light that has passed through a repeatedly scattering medium can provide visualization of the spatial distribution of inhomogeneities of this medium [9–11].

The use of this method is based on modulation of the polarization azimuth, probing laser radiation and detection of changes in the polarization parameters of light transmitted through an anisotropic biological tissue material.

One of the well-known techniques that allows you to obtain information about the polarization properties of tissues and identify differences between healthy and tumor tissue is called Mueller polarimetric imaging. For example, in the study of colon cancer, it was shown that the malignant zone has a lower degree of depolarization compared to the surrounding healthy tissue [12, 13].

In addition, the optical characteristics of internal tissues obtained by polarimetry in the visible and near infrared ranges of the spectrum contain a large amount of morphological and functional information of potential biomedical significance. This is especially important when using the polarization-optical method in combination with neuroimaging techniques to observe changes in human brain activity in real time [14, 15].

The integration of the polarization-optical method with SEMS will allow obtaining high-level images of the sample surface with very high resolution, which is especially useful for studying nanoscale details of the structure of tissues and tumors.

At the moment, the clinical application of polarized light in the diagnosis of oncology is at an early stage. Additional research is needed to confirm the effectiveness of polarization imaging in the diagnosis, staging and monitoring of cancer. In this regard, the purpose of this work is to investigate the possibility of differentiation of pathological tissues by the method of polarization imaging and digital processing of recorded images.

2 The Research Method

As a result of the interaction of polarized light with an absorbing anisotropic biological tissue, various polarization effects arise, such as linear delay (phase difference between two orthogonal linear polarizations), circular delay or optical rotation (phase difference between right and left light with circular polarization), linear and circular attenuation (differential attenuation of orthogonal polarizations as for the state of linear and for circular polarization) and partial depolarization. All these polarization effects occur simultaneously, rather than in an orderly or sequential manner, which makes it difficult to interpret them unambiguously [16, 17].

The extraction of information about the polarization properties of the medium, regardless of any assumptions about the order of their occurrence, is possible using a Mueller polarimeter, which makes it possible to unambiguously interpret the polarization interactions [16, 18]. These polarization effects can be used to study the composition and microstructure of the sample under study.

The generalized scheme of the Mueller polarimeter consists of an optical radiation source; a polarization state generator from a linear polarizer and a quarter-wave phase plate; an object under study; a polarization state analyzer from a quarter-wave phase plate and a linear polarizer; a photodetector [19, 20].

When the Stokes vector $\mathbf{S}_{\text{in}}$ of incident light propagates through a scattering medium, the output Stokes vector $\mathbf{S}_{\text{out}}$ can be obtained as a linear combination of four parameters of the Stokes vector of incident light [18]:

$$\mathbf{S}_{\text{out}} = \mathbf{M} \cdot \mathbf{S}_{\text{in}} \tag{1}$$

or in the following matrix form [18]:

$$\begin{bmatrix} S_{1,\text{out}} \\ S_{2,\text{out}} \\ S_{3,\text{out}} \\ S_{4,\text{out}} \end{bmatrix} = \begin{bmatrix} M_{11} & M_{12} & M_{13} & M_{14} \\ M_{21} & M_{22} & M_{23} & M_{24} \\ M_{31} & M_{32} & M_{33} & M_{34} \\ M_{41} & M_{42} & M_{43} & M_{44} \end{bmatrix} \cdot \begin{bmatrix} S_{1,\text{in}} \\ S_{2,\text{in}} \\ S_{3,\text{in}} \\ S_{4,\text{in}} \end{bmatrix}, \tag{2}$$

where $\mathbf{M}$ is the Mueller matrix of the sample under study, and M_{ij} are the experimentally measured elements of the matrix, $i, j = 1 \ldots 4$.

The elements of the Mueller matrix can be calculated based on measuring the radiation flux at the output of the polarimeter with six combinations of optical components of the polarization state generator and analyzer: H is linear horizontal polarization; V is linear vertical polarization; P is linear polarization 45°; Q is linear polarization – 45°; R and L are circular right and left polarizations, respectively. Thus, the Mueller matrix in expression (2) can be transformed to the form [21]:

$$\mathbf{M} = \frac{1}{4}\begin{bmatrix} HH + HV + VH + VV & HH + HV - VH - VV & PH + PV - QH - QV & RH + RV - LH - LV \\ HH - HV + VH - VV & HH - HV - VH + VV & PH + QV - QH - PV & RH + LV - LH - RV \\ HP - HQ + VP - VQ & HP + VQ - VP - HQ & PP + QQ - QP - PQ & RP + LQ - LP - RQ \\ HR - HL + VR - VL & VL + HR - HL - VR & QL + PR - PL - QR & LL + RR - RL - LR \end{bmatrix}. \tag{3}$$

In Eq. (3), each element of the Mueller matrix is calculated from four measurements at different states of supply and receiving polarization. For example, *HV* means that the linear vertical polarization component is analyzed taking into account the incidence of linearly horizontally polarized light.

In practice, the use of the full Mueller matrix is difficult, since the azimuth dependence of most matrix elements (12 out of 16) changes when the spatial position of the object of study changes relative to the axis of incidence of radiation. Azimuthally stable, that is, those that do not depend on the angle of rotation of the object of study, are the elements of the Mueller matrix: $M_{11}, M_{14}, M_{41}, M_{44}$ [22, 23].

When normalizing the Mueller matrix, the first element M_{11} is always equal to one, and the fourth element M_{44} contains information about the radiation parameters at the sample output when its input is exposed to circularly polarized radiation.

Therefore, to study the polarization properties of biological tissues, it is sufficient to determine three elements of the Mueller matrix M_{14}, M_{41}, M_{44}, which characterize the processes of converting the medium under study from circularly polarized radiation to linearly polarized and converting linearly polarized radiation to circularly polarized, respectively.

The use of matrix photodetectors with a single field of analysis in the Mueller polarimeter scheme makes it possible not only to visualize the results, but also to objectively assess the features of the obtained coordinate distributions of irradiation in images formed by radiation after its interaction with pathological tissues [24].

Statistical analysis of the spatial distributions of irradiance corresponding to various elements of the Mueller matrix of the sample will allow to identify and evaluate statistically significant differences between the study group.

First of all, it is necessary to determine the sample size, that is, the number of elements of the analyzed radiation distribution that will be studied. To do this, the scattering spot registered on the matrix photodetector is divided into a certain number of elementary cells corresponding to the number of pixels of the matrix. Information about the irradiation of each of the cells is considered separately from each other. A larger sample has greater statistical power and allows you to get more reliable results. Then, a set of statistical moments from the 1-st to the 4-th moments are calculated [23]:

$$Z_1 = \frac{1}{N} \cdot \sum_{j=1}^{N} q_j; \; Z_2 = \sqrt{\frac{1}{N-1} \cdot \sum_{j=1}^{N} (q_j - Z_1)^2};$$

$$Z_3 = \frac{1}{Z_2^3} \cdot \frac{1}{N} \cdot \sum_{j=1}^{N} (q_j - Z_1)^3; \; Z_4 = \frac{1}{Z_2^4} \cdot \frac{1}{N} \cdot \sum_{j=1}^{N} (q_j - Z_1)^4, \tag{4}$$

where N is the number of elements in the area of interest; q_j is the numerical value of the parameter in the cell; Z_1 is the mathematical expectation; Z_2 is the standard deviation; Z_3 is the skewness; Z_4 is the kurtosis.

Thus, it was shown that the Mueller method, in combination with statistical techniques for processing the obtained distributions, is highly informative and can be used for laboratory studies of samples with a complex anisotropic structure.

3 Experimental Research

The purpose of this experimental study is to demonstrate the capabilities of the polarization imaging method with subsequent digital image processing to differentiate biological samples with pathological tissues. Measurements of spatial distributions of irradiation on photosensitive sites of multi-element photodetectors and further calculation of the elements of the Mueller matrix for each sample of pathological tissue were carried out.

The experiment was carried out using a four-channel robotic polarimetric system with multi-element imaging sensors designed for use as part of a diagnostic complex based on SEMS modules [25].

The Mueller matrices of biological tissue samples with pathological changes caused by adenocarcinoma were measured. Biopsy samples of the stomach and lung were examined, representing sections of a small part of the organ 0.5 mm thick, 3 mm in diameter. To preserve the integrity of the tissue structure, the samples are placed in a special solution that helps to keep the biological object in an acceptable condition for longer. Based on the measurements, a comparative analysis of the dependencies of the optical parameters of tissues of different organs with various pathological conditions was performed.

Each object has its own degree of differentiation, which makes it possible to classify adenocarcinoma according to the degree of maturity of tumor cells. This makes it possible to more accurately assess the aggressiveness of the tumor and predict its behavior, which is important for choosing the optimal treatment and predicting the outcome of the disease.

In this regard, the degree of differentiation is divided into four types: G1 is a highly differentiated adenocarcinoma; G2 is a moderately differentiated adenocarcinoma; G3 is a low–differentiated adenocarcinoma; G4 is an undifferentiated adenocarcinoma.

In this experimental study, gastric adenocarcinoma with a moderate (G2) degree of differentiation, as well as low-grade (G3) lung metastasis were considered (Table 1).

Table 1 Samples for the study

Sample	Characteristic	Types	Features
248/AC	Gastric adenocarcinoma	G2	The presence of atypical cells similar to low-grade invasive cancer
248/MTS	Lung metastasis	G3	The cells are very different from normal ones and have morphological structures characteristic of cancer

The methodology of the experiment consists of four stages: stage I is the preparation of samples; stage II is the preparation of the components of the experimental setup; stage III is the measurement of the distribution of illumination on the sample at different positions of the axes of the polarization elements; stage IV is the preservation and processing of the results. Let's consider the stages presented sequentially.

Stage I. Preparation of samples

Before starting measurements, preliminary preparation of the sample should be carried out, namely, blot it on filter paper in order to get rid of excess moisture of the solution. After that, the object is placed in the plane of the experimental setup's slide table between two plane-parallel plates in order to avoid drying out. Plane-parallel plates either transmit or reflect the input radiation and, thus, do not affect the parameters of the radiation scattered at an angle by the sample under study. The recommended room temperature during the experiment is 20–22 °C.

Stage II. Preparation of the components of the experimental setup

The polarization elements of the optical circuit of the measuring system need to be adjusted.

Initially, the polarizer and analyzer are positioned so that there is a complete dimming of the light at the output of the system. This position corresponds to a 90° angle between the anisotropic axes (crossed position). Accordingly, when the analyzer is rotated by 90°, it will be possible to achieve a consistent position (0°). Between these positions, the rotation of the anisotropic axis of the analyzer by 45° is fixed, on the other side of the zero mark – 45° is fixed.

After successful adjustment of the specified elements of the polarization filters in the generator and the polarization state analyzer of the experimental installation, it is necessary to correctly place the quarter-wave phase plate in the polarization state analyzer: its axis should be located at 45° to the transmission axis of the analyzer. To find this position, at the moment of rotation of the axis of the analyzer at 45°, a quarter-wave phase plate is introduced into the course of the rays, and by gradual rotation of this plate, such an arrangement of its axis is achieved in which the introduction of an additional auxiliary polarizer after it does not affect the intensity of the transmitted radiation. When the plate is placed correctly, it converts linearly polarized radiation into circularly polarized radiation. The quarter-wave phase plate in the polarization state generator is used to generate circularly polarized radiation.

Before starting work, make sure that the components of the experimental installation are securely attached. In case of the slightest shift of any of the components, further results can be considered incorrect. In this regard, to account for the relative mutual displacements of images on the photodetector site, a program for correcting the position of images based on the alignment of their energy centers was developed and implemented [26].

The results of the polarization-optical analysis are influenced by extraneous radiation sources. To avoid their influence, measurements should be carried out in complete darkness, with a light-tight casing.

Stage III. Measurement of the distribution of illumination on the sample at different positions of the axes of the polarization elements

The experimental technique includes measuring irradiance on a matrix photodetector six times: for the first four in the experimental setup, only the axes of the polarizer and analyzer in the polarization state generator and analyzer are rotated, for the last two measurements, in addition, a phase-shifting element (a quarter-wave phase plate in the polarization state analyzer) is introduced into the course of the rays. Each time, the intensities of the waves that have passed through the analyzer located in the path of this wave and transmitting radiation with azimuths are recorded.

Thus, during polarization measurements, part of the outgoing radiation is reflected from the biological tissue, part is scattered, and the remaining part is absorbed. The presence of scattering indicates the degree of depolarization, and the percentage of reflected radiation can be estimated using the parameters of the polarized component of the radiation (azimuth and ellipticity of the polarized radiation).

In order to obtain 16 elements of the Mueller matrix, at least 16 types of intensity measurements are required. It is necessary to perform 16 different combinations of the generator and the polarization state analyzer. This is achieved by rotating the elements of the generator and analyzer of the polarization system at the desired angles: the transmission planes of the polarizer and analyzer are located at an angle of 45° when the analyzer is rotated clockwise from the zero position, a quarter-wave phase plate is introduced into the course of the rays; the transmission planes of the polarizer and analyzer are located at an angle of 45° when the analyzer is rotated counterclockwise from the zero position, a quarter-wave phase plate is introduced into the course of the rays.

Stage IV. Preservation and processing of the results

A total of 36 measurements were made for the samples, corresponding to each variation in the orientation of the anisotropic axes of the polarization elements of the experimental installation.

After receiving the results of the initial measurements, each one should be processed. The developed software was used to measure and store data on the distribution of irradiance at the sensitive site of the matrix photodetector along the red channel, which was presented in [26].

To prevent inaccuracies and to obtain the most reliable results, it was necessary to centralize all images in relation to the first dimension. The final stage of the image processing part of the analysis was the assignment of a zero light intensity value outside the area of interest. Thus, after the processing stage, correct arrays were obtained at the output, which were used to form linear combinations corresponding to each of these matrix elements according to formula (3). As a result, coordinate distributions were obtained, collectively forming the Mueller matrix of the sample under study.

For further analysis of the polarization properties of pathological samples based on statistical methods of processing the results, the elements of the Mueller matrix

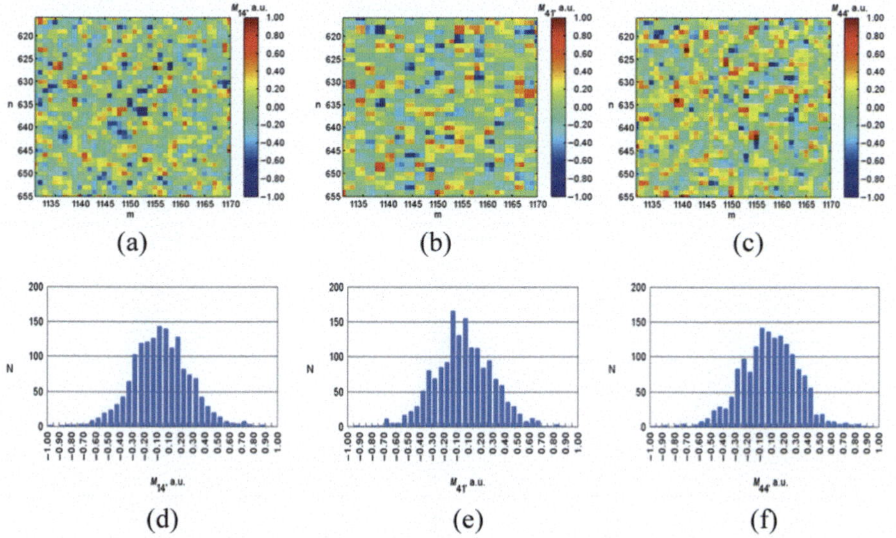

Fig. 1 Coordinate distributions and histograms of Mueller matrix invariants: M_{14} (**a**, **d**), M_{41} (**b**, **e**), M_{44} (**c**, **f**) for a sample of gastric adenocarcinoma; N is the frequency of values in the allocated area $(m \times n)$ of the matrix photodetector

of the sample M_{14}, M_{41}, and M_{44}, invariant to its spatial orientation, as well as their coordinate distributions were calculated.

4 Experimental Results

The work on processing and visualizing the results was carried out using the MATLAB software package in accordance with the previously discussed methodology. The obtained coordinate distributions and histograms of the Mueller matrix invariants for the sample of gastric adenocarcinoma and lung metastasis are shown in Figs. 1 and 2, respectively.

The statistical parameters used to characterize the depolarization distribution $\Delta(m \times n)$ of these samples are presented in Tables 2 and 3, respectively.

5 Discussion

In all considered coordinate distributions, the second moments exceed the corresponding first moments (Fig. 1a, b, c and Fig. 2a, b, c). This is due to the significant variation of the distribution options in both positive and negative areas of numerical values. The variance of the distribution of values increases Δ monotonously with an

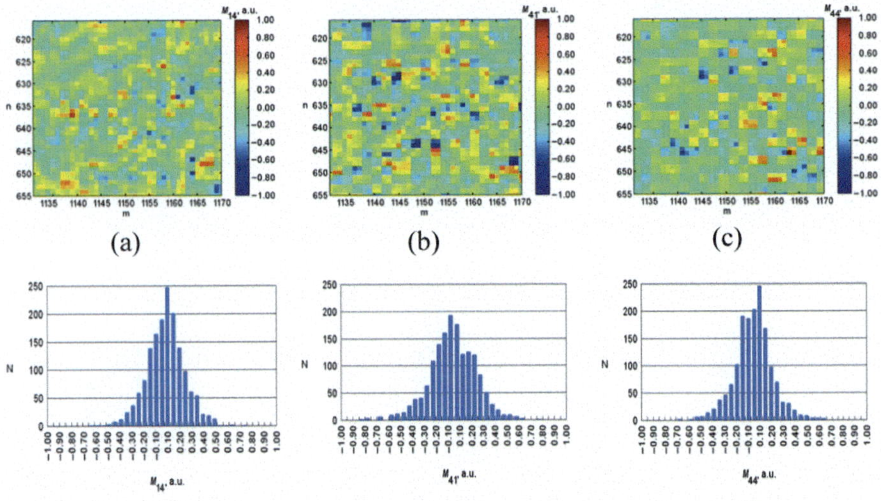

(a) (b) (c)

(d) (e) (f)

Fig. 2 Coordinate distributions and histograms of Mueller matrix invariants: M_{14} (**a**, **d**), M_{41} (**b**, **e**), M_{44} (**c**, **f**) for a sample of lung metastasis; N is the frequency of values in the allocated area $(m \times n)$ of the matrix photodetector

Table 2 Statistical moments from the 1-st to the 4-th moments of distributions of the Mueller matrix invariants of the gastric adenocarcinoma sample

Statistical moment / The invariant	Z_1	Z_2	Z_3	Z_4
M_{14}	– 0.016	0.26	– 0.073	4.23
M_{41}	0.004	0.26	– 0.29	4.55
M_{44}	0.042	0.27	0.025	4.55

Table 3 Statistical moments from the 1-st to the 4-th moments of distributions of the Mueller matrix invariants of the lung metastasis sample

Statistical moment / The invariant	Z_1	Z_2	Z_3	Z_4
M_{14}	0.0098	0.17	0.019	4.68
M_{41}	– 0.0015	0.23	– 0.71	6.14
M_{44}	– 0.025	0.18	0.17	4.94

increase in the average value of depolarization. On the other hand, since the samples are represented by slightly different structures and with their degree of differentiation, they are characterized by an "individual" half-width, asymmetry and sharpness of the peak of the $N(\Delta)$ distribution.

If we look from the point of view of biophysics, then the state of "cancer" is characterized by an increase in birefringence of protein structures. That is, the following analytical script is implemented for malignant neoplasms: $\uparrow M_{41} \Leftrightarrow M_{14} \downarrow$.

Anisotropic samples do not have a distinct fibrous, fibrillar structure. Consequently, the component of their images, which is heterogeneous in polarization, is formed by the values of azimuths and ellipticals.

The invariants of the matrix of the surface structure of the diffuser (M_{14}, M_{41}, M_{44}) provide azimuthally stable information about the optically anisotropic properties of the object under study. As you can see, the depolarization maps of objects are spatially heterogeneous. This fact may be related to the coordinate heterogeneity of the sample structure. Optically, this is reflected in fluctuations in the coordinates of the depolarization degree values. As the multiplicity dispersion increases, these fluctuations decrease. The coordinate distribution of the depolarization parameter Δ for this sample is characterized by a wider range of values. As a result, there is a gradual decline in dependence. This is reflected in the histograms (Fig. 1d, e, f and Fig. 2d, e, f).

As it was said, the presence of a pathological process in the tissue is characterized by the formation of a system of "elliptical" emitters and an increase in birefringence of the fibrillar network of adenocarcinoma. Consequently, there is a shift of the main extremum of the histogram of coordinate distributions N. As a result, mathematical expectation (Z_1), skewness (Z_3) and kurtosis (Z_4) increase significantly. The analysis of the coordinate distributions of the invariants M_{14}, which characterizes the processes of converting circularly polarized radiation into linearly polarized, reveals a shift of the main extremum of the histogram $N(M_{14})$ to a region of lower values in the plane of the histological section of adenocarcinoma tissue. The reverse pattern is observed for the distributions of the Mueller matrix invariant characterizing the processes of converting linearly polarized radiation into circularly polarized radiation. The histological section of metastasis is characterized by an increase in the probability of extreme values of the invariant M_{41}.

The analysis of the obtained results showed that the following parameters are the most sensitive to an increase in the multiplicity of scattering in the phantom volume: dispersion and the statistical moment of the 4-th moment (kurtosis), which characterizes the sharpness of the peak distribution of the depolarization parameter values, is practically static for the case with a degree of G2, but has changes of about 1.3 times for G3.

Thus, the high sensitivity of a set of statistical parameters to changes in the coordinate distributions of depolarization Δ of samples with different scattering coefficients, and therefore different degrees of differentiation, is confirmed.

These findings open up prospects for evaluating the quantitative characteristics of optically anisotropic objects, such as biological tissues of other organs with different types and degrees of malignancy of neoplasms. These samples represent much more complex structures compared to those considered during this experiment.

6 Conclusion

As a result of the experimental study, azimuthally stable invariants of the Mueller matrix were determined for samples of biological tissue of different organs and their degree of malignancy. It is proved that they characterize the polarization manifestations of porphines against the background of linear birefringence and optical activity of proteins of biological tissues. A study of the diagnostic capabilities of the Mueller matrix polarimetry method was conducted in the comparative analysis of biopsy samples of gastric adenocarcinoma and lung metastasis with varying degrees of malignancy of the neoplasm. A change in the structure of the tissue due to its pathology also changes its optical properties and increases the scattering of radiation.

Thus, further studies of the degree of malignancy of biological tissue samples at different stages of pathology development will allow us to develop a database of optical indicators for their differentiation.

The polarization imaging method based on the use of information about the elements of the Mueller matrix of samples demonstrates the potential for screening optical anisotropy in biological tissue, including for the diagnosis of cancerous tissues.

At the same time, the development and implementation of polarization systems in diagnostic complexes based on SEMS modules can significantly improve the capabilities of diagnosis and tracking of cancerous tissues. The development of polarization systems based on SEMS modules seems to be a promising continuation of this research in order to introduce such devices into clinical practice for the diagnosis of cancerous tissues.

Thus, the integration of the polarization method with SEMS opens up new opportunities for the diagnosis and monitoring of oncological diseases. However, its practical application requires further research, standardization and development of appropriate technologies.

Acknowledgements This research was funded by the Ministry of Science and Higher Education of the Russian Federation under the agreement № 075-15-2022-303 to support the development of a World-class research center "Pavlov Center for Integrative Physiology for Medicine, High-tech Healthcare, and Stress Tolerance Technologies."

References

1. Crosby, D., Bhatia, S., Brindle, K.M., Coussens, L.M., Dive, C., Emberton, M., Esener, S., Fitzgerald, R., Gambhir, S.S., Kuhn, P., Rebbeck, T., Balasubramanian, S.: Early detection of cancer. Science **375**(6586), eaay9040 (2022)
2. Al-Shaheri, F.N., Alhamdani, M.S.S., Bauer, A.S., Giese, N., Büchler, M.W., Hackert, T., Hoheisel, J.D.: Blood biomarkers for differential diagnosis and early detection of pancreatic cancer. Cancer Treat. Rev. **96**, 102193 (2021)

3. Nasery, M.M., Varzandeh, M., Pahlavanneshan, S., Mohamadi, N., Sarhadi, S., Fekri, H.S., Mahammadinejad, R., Ahn, K.S.: Curcumin: a potential therapeutic natural product for adenocarcinomas. Phytochem. Lett. **49**, 45–55 (2022)
4. Shahriari, N., Grant-Kels, J.M., Rabinovitz, H., Oliviero, M., Scope, A.: Reflectance confocal microscopy: diagnostic criteria of common benign and malignant neoplasms, dermoscopic and histopathologic correlates of key confocal criteria, and diagnostic algorithms. J. Am. Acad. Dermatol. **84**(1), 17–31 (2021)
5. Chugh, V., Basu, A., Kaushik, A., Manshu, Bhansali, S., Basu, A.K.: Employing nano-enabled artificial intelligence (AI)-based smart technologies for prediction, screening, and detection of cancer. Nanoscale (2024)
6. Zhang, S., Wei, S., Liu, Z., Li, T., Huang, X.L., Wang, C., Xie, Z., Al-Hartomy, O.A., Al-Ghamdi, A.A., Wageh, S., Gao, J., Tang, Y., Wang, H., Wang, Q., Zhang, H.: The rise of AI optoelectronic sensors: from nanomaterial synthesis, device design to practical application. Mater. Today Phys., 100812 (2022)
7. Lee, H.E.: Novel bio-optoelectronics enabled by flexible micro light-emitting diodes. Electronics **10**(21), 2644 (2021)
8. Naghdi, T., Ardalan, S., Adib, Z.A., Sharifi, A.R., Golmohammadi, H.: Moving toward smart biomedical sensing. Biosens. Bioelectron. **223**, 115009 (2023)
9. Ávila, F.J., Bueno, J.M.: Polarization Techniques in Biological Microscopy. Optical Polarimetric Modalities for Biomedical Research, pp. 27–45. Springer International Publishing, Cham (2023)
10. He, C., He, H., Chang, J., Chen, B., Ma, H., Booth, M.: Polarisation optics for biomedical and clinical applications: a review. Light. Sci. Appl. **10**(1), 194 (2021)
11. Li, X., Han, Y., Wang, H., Liu, T., Chen, S.C., Hu, H.: Polarimetric imaging through scattering media: a review. Front. Phys. **10**, 815296 (2022)
12. Le, D.L., Nguyen, D.T., Le, T.H., Phan, Q.H., Pham, T.T.H.: Characterization of healthy and cancerous human skin tissue utilizing Stokes–Mueller polarimetry technique. Opt. Commun. **480**, 126460 (2021)
13. Singh, M.D., Ghosh, N., Vitkin, I.A.: Mueller matrix polarimetry in biomedicine: enabling technology, biomedical applications, and future prospects. In: Polarized Light in Biomedical Imaging and Sensing: Clinical and Preclinical Applications, pp. 61–103. Springer International Publishing, Cham (2022)
14. Yen, C., Lin, C.L., Chiang, M.C.: Exploring the frontiers of neuroimaging: a review of recent advances in understanding brain functioning and disorders. Life **13**(7), 1472 (2023)
15. Novikova, T., Pierangelo, A., Schucht, P., Meglinski, I., Rodríguez-Núñez, O., Lee, H.R.: Mueller polarimetry of brain tissues. In: Polarized Light in Biomedical Imaging and Sensing: Clinical and Preclinical Applications, pp. 205–229. Springer International Publishing, Cham (2022)
16. Li, P., Dong, Y., Wan, J., He, H., Aziz, T., Ma, H.: Polaromics: deriving polarization parameters from a Mueller matrix for quantitative characterization of biomedical specimen. J. Phys. D Appl. Phys. **55**(3), 034002 (2021)
17. Shi, Y., Chen, B., He, H., Ma, H.: Quantitative assessment of tissue structures based on Mueller matrix polarimetry and derived parameters imaging. In: Polarized Light and Optical Angular Momentum for Biomedical Diagnostics 2022, SPIE, vol. 11963, pp. 77–86 (2022)
18. Gil, J.J., Ossikovski, R.: Polarized Light and the Mueller Matrix Approach. CRC Press (2022)
19. Arteaga, O., Bian, S.: Mueller matrix imaging. Opt. Polarim. Modalities Biomed. Res., 77–99 (2023)
20. Li, J., Liao, R., Guan, C., Wang, H., Zhuo, Z., Zeng, Y., Ma, H.: Particulate Mueller matrix polarimetry. Opt. Laser Technol. **158**, 108780 (2023)
21. Wang, X., Lai, J., Li, Z.: Polarization studies for backscattering of RBC suspensions based on Mueller matrix decomposition. Opt. Express **20**(18), 20771–20782 (2012)
22. Grytsyuk, M., Tomka, Y., Gorsky, M., Soltys, I., Talakh, M., Drin, Y., Yatsko, O., Gurina, O., Garazdyuk, M., Litvinenko, O., Dubolazov, O.: Muller-matrix invariants of linear and circular birefringence of polycrystalline films of biological liquids pathologically and necrotic changed human bodies. In: Biosensing and Nanomedicine XII, SPIE, vol. 11087, pp. 25–32 (2019)

23. Ushenko, A.G., Dubolazov, A.V., Ushenko, Y.A., Kushnerick, L.Y., Saknovskiy, M.Y., Zhytaryuk, V.G., Lacusta, I., Prydiy, O.G., Grygorishin, P.: Differentiation of Mueller-matrix invariants of biological tissues fibrillar networks with different phase and amplitude anisotropy. Semicond. Phys. Quantum Electron. Optoelectron. **20**(1), 129–136 (2017)
24. Koestner, D., Stramski, D., Reynolds, R.A.: Polarized light scattering measurements as a means to characterize particle size and composition of natural assemblages of marine particles: erratum. Appl. Opt. **60**(2), 380–382 (2021)
25. Khlynov, R.D., Ryzhova, V.A., Konyakhin, I.A., Korotaev, V.V.: Robotic polarimetry system based on image sensors for monitoring the rheological properties of blood in emergency situations. In: Smart Electromechanical Systems: Recognition, Identification, Modeling, Measurement Systems, Sensors, pp. 201–218. Springer International Publishing, Cham (2022)
26. Khlynov, R.D., Ryzhova, V.A., Yarishev, S.N., Konyakhin, I.A., Korotaev, V.V., Shelepin, Y.E., Djamiykov, T.S., Marinov, M.B.: Analysis of polarization images in the microphysical blood parameters research for the hematocrit diagnostics. Micromachines **13**(2), 2241 (2022)

Comparative Analysis of Methods for Classifying Electroencephalography Signals for a Robotic Brain-Computer Interface

Artemii Y. Zhdanov, Victoria A. Ryzhova, Ruslan D. Khlynov, Igor A. Konyakhin, and Yuri E. Shelepin

Abstract *Problem statement*: The implementation of a brain-computer interface for robotic systems based on motor activity is carried out by classifying brain activity signals. Electroencephalography (EEG) is the most commonly used method to measure brain activity due to its mobility and high temporal resolution. Currently, extracting meaningful features from a large volume of EEG data in the presence of noise, followed by signal classification, is not a solved problem. *Purpose of research*: To determine the accuracy of various classification methods for obtaining information about motor activity from EEG data. *Method*: The performance of classification methods is assessed using the BCI competition IV-dataset 1, which contains EEG data obtained from 9 people performing motor activity tasks. Pre-processing of EEG data consists of finding the spectrum of the EEG signal using the Fast Fourier Transform (FFT) algorithm. The classification methods are implemented in Python. The libraries numpy, tensorflow, keras, sklearn, PIL are used in processing EEG data. *Results*: The performance of such signal classification methods as Support Vector Machine, k-Nearest Neighbors method, Linear Discriminatory Analysis, Decision Tree Classifier, Convolutional Neural Networks. *Practical significance*: The information about the performance of various approaches to the classification of brain activity signals makes it possible to create effective systems of brain-computer interfaces for robotic systems.

A. Y. Zhdanov (✉) · V. A. Ryzhova · R. D. Khlynov · I. A. Konyakhin
ITMO University, Saint-Petersburg, Russia
e-mail: artemichzdanov@gmail.com

V. A. Ryzhova
e-mail: victoria_ryz@itmo.ru

R. D. Khlynov
e-mail: ruslan384ruslan3@gmail.com

I. A. Konyakhin
e-mail: iakoniakhin@itmo.ru

A. Y. Zhdanov · Y. E. Shelepin
Pavlov Institute of Physiology, Russian Academy of Sciences, Saint-Petersburg, Russia
e-mail: shelepin-yuri@infran.ru

I. L. Tarasova and B. A. Kulik (eds.), *Smart Electromechanical Systems*, Studies in Systems, Decision and Control 544, https://doi.org/10.1007/978-3-031-64277-7_20

Keywords BCI · Robotic systems · SVM · k-Nearest neighbour · MLP · EEG · LDA · CNN · DTC

1 Introduction

Today, research is actively underway in the field of creating tools and methods for direct interaction between the human brain and a computer [1, 2]. The set of such tools and methods is called a brain-computer interface. In the context of robotic systems, brain computer interface allows the creation of robotic systems controlled by signals from brain activity [2]. Accordingly, the task of a robotic brain-computer interface is to convert brain activity signals into control signals of a robotic system. Various neuroimaging techniques are used to obtain signals of brain activity. The most common method of neuroimaging in the implementation of a brain-computer interface is the method of electroencephalography (EEG), due to its high mobility, relative cheapness and high time resolution [3].

The principle of generating control signals based on brain activity signals is the classification of EEG data. The choice of a classifier for brain activity signals largely depends on the characteristics by which the classification will be performed, as well as the number of classes. Comparison of the effectiveness of classification approaches is carried out in most cases using one of the well-known metrics, most often using the "F1 score" metric [4, 5]. Studying the effectiveness of various approaches to classifying brain signals makes it possible to optimize the functioning of brain-computer interfaces as part of robotic systems [5].

Thus, the purpose of this study is to perform the comparative analysis of the performances of various classification methods for obtaining information about motor activity from EEG data.

2 Data Description

To assess the accuracy of various classifiers, the "BCI Competition IV dataset 1" data set was used [6]. The data set includes EEG data taken while imagining motor activity of one of the hands, that is, the subject was imagining the movement of the right or left hand. The captured EEG has 59 channels. Data were collected with visual stimulation indicating movement of the left or right hand using arrows indicating the direction on the screen. The subject was given 4 s to complete this task. These periods alternated with two-second periods of a blank screen and two-second periods of a fixation cross being displayed on the screen. Each subject completed a total of 200 tasks. In the data set, each of the tasks is marked as belonging to one of two classes "left" or "right". The dataset contains EEG recordings from four healthy subjects and three artificially generated recordings. Each entry is labeled with the letter "a"—"g".

In this work, data marked with the letter "d" was used. The data was split 80/20 into training data and validation data.

3 Feature Extraction

When using the EEG method in brain-computer interface systems as a neuroimaging method, it becomes necessary to extract informative features from the entire EEG data set. Extraction of informative features from a raw EEG data set occurs using a sequence of mathematical processing algorithms. The extracted features are then fed to the classifier input. The quality of the classifier depends on the selected features, so choosing an appropriate feature extraction method is important [7].

Motor activity (or its imagining) is characterized by suppression of signals from brain activity in the region of μ-rhythms [8]. Therefore, the required information features—that is, signs of signal suppression in the region of μ-rhythms—are located in the spectral region.

Initially, the EEG data array was a matrix of 59 EEG channels with 190,473 samples each. To extract features, the volume of EEG data pertaining to one subject was divided into 200 parts (hereinafter referred to as trials), one part for each completed task. Each trial lasted 2 s and was divided into 100 samples. Time periods not relevant to testing were discarded. The EEG signal related to the passage of each test was converted from the time domain to the frequency domain by calculating the power spectral density (hereinafter referred to as PSD) of the test. Figure 1 shows a plot of the power spectral density of the signal at electrode C4 during one of the tests.

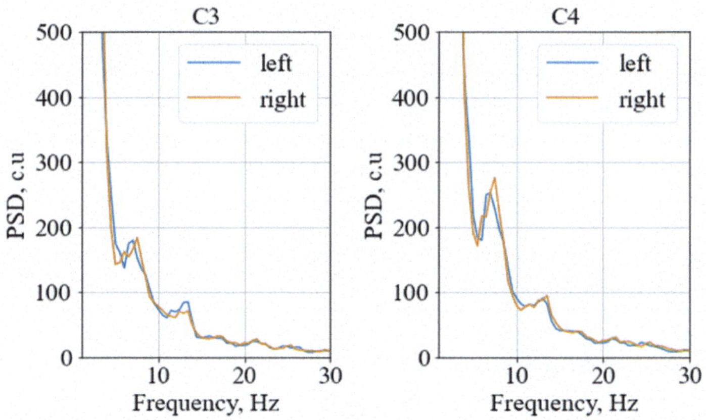

Fig. 1 PSD of the signal

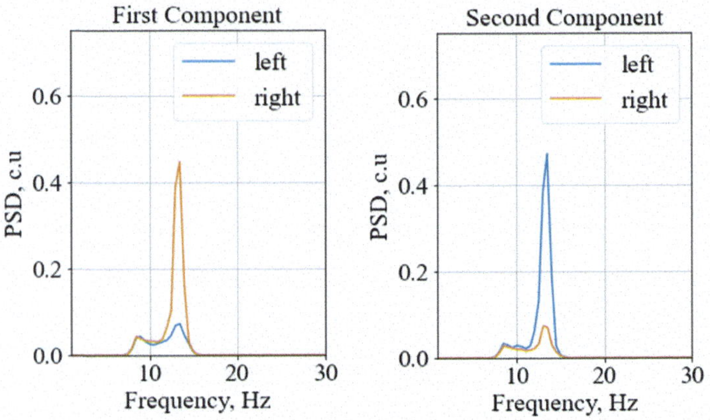

Fig. 2 Filtered PSD

Further preprocessing consisted of filtering the obtained power spectral density of the EEG signal during the test using a bandpass filter with a passband of 8–15 Hz. The choice of filter bandwidth is determined by the spectrum of signals of sensorimotor activity of the human brain. In addition to frequency filtering, in order to reduce the dimension of the feature space, a spatial filtering algorithm (Common Spatial Pattern Filter, abbreviated CSP) is applied to the data [9]. CSP, based on these channels, generates its own components for classification, which allows you to select for classification only those channels of data components that have the maximum difference between each other.

The application of the CSP algorithm to the obtained data made it possible to reduce the number of analyzed components, that is, the dimension of the feature space, to two. Figure 2 shows the component values during the same test.

After filtering the resulting power spectral density, in accordance with the feature extraction method proposed in [10], the natural logarithms of the variances of the power spectral densities related to one test are used as features supplied to the input of the classifiers. Thus, one number is used as a sign for further classification—the natural logarithm of the variance of the filtered PSD.

4 EEG Signal Classification

4.1 K-Nearest Neighbors

The k-Nearest Neighbor algorithm [11] is a nonparametric algorithm used for classification. The k-NN algorithm is widely used in supervised learning problems and has been well studied in the field of machine learning. A nearest neighbor classifier assigns a feature to some class based on its nearest neighbors in the feature space. The algorithm was implemented using the "scikit-learn" library.

4.2 *Support Vector Machine*

Support Vector Machine (SVM) [12] is a parametric algorithm used to solve classification and regression problems. The main idea of the method is to construct a hyperplane that separates the sample objects in such a way that the distance between the separating hyperplane and the separated objects becomes the greatest possible. The regu parameter in this algorithm is the kernel function. Since the task is a binary classification of clearly geometrically separated states of brain activity—representing the motor activity of the right or left hand—a linear function was chosen as the kernel function. The algorithm was implemented using the "scikit-learn" library.

4.3 *Linear Discriminatory Analysis*

Linear Discriminatory Analysis (LDA) [13], also known as Normal Discriminatory Analysis (NDA) is a nonparametric algorithm that is a generalization of Fisher's linear discriminant, a method used in statistics and other fields—with the goal of finding a linear combination of features that characterizes or separates two or more classes of objects or events. The resulting combination can be used as a linear classifier. The LDA algorithm seeks to find those linear combinations of features that best separate the classes, minimizing the variation within each class and maximizing the variation between classes. The resulting linear functions can be used to classify new data, helping to identify which class it belongs to. The algorithm was implemented using the "scikit-learn" library.

4.4 *Decision Tree Classifier*

Decision Tree (DT) [14] is a parametric algorithm used for classification and regression. The main task is to create a model that predicts the value (class) of some variable based on decision rules derived from the characteristics of the data. In the context of classification, a decision tree divides data into subgroups based on feature values, thereby creating a tree-like structure. This allows you to determine whether objects belong to different classes based on their characteristics. For this classifier, the adjustable parameter is the selection criterion. Since in this case the classification problem is being solved, the Gini criterion was used as a selection criterion. The algorithm was implemented using the "scikit-learn" library.

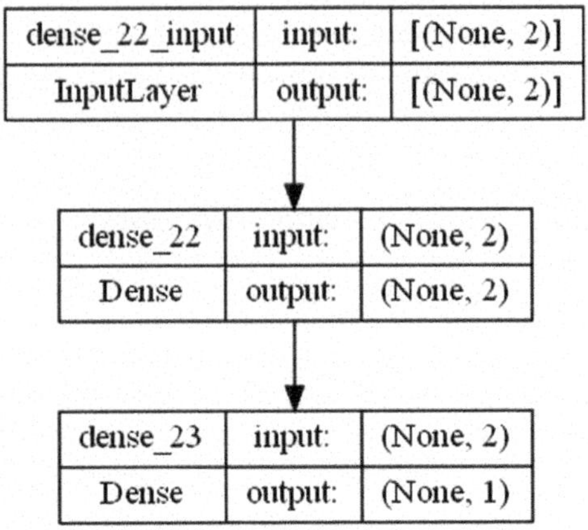

Fig. 3 Architecture of the synthesized MLP-NN

4.5 Multi-layer Perceptron Neural Network

Multi-Layer Perceptron Neural Network (MLP-NN) [15] is a widely used neural network structure consisting of three main layers: input layer, hidden layer and output layer. This neural network is trained using the backpropagation algorithm. The construction of the neural network was implemented using the "keras" library. The architecture of the neural network used in the current work is presented in Fig. 3.

4.6 Convolutional Neural Network

The essence of the concept of convolutional neural networks (English Convolutional Neuron Network, abbreviated CNN) is the alternation of convolutional layers and subsampling layers [16]. The network architecture is a unidirectional structure that does not contain feedback, and has a multi-layer organization. To train such networks, the backpropagation method is used.

In this study the CNN classifier will be given spectrogram data without CSP filtration. This will be done to compare the manually done data preprocessing used in previous classifiers and automatic preprocessing that will be achieved through learning done by the neural network.

The neural network is synthesized using the 'keras' library. The synthesized network has ten layers. Convolutional layers are interspersed with dimensionality reduction layers. The neuron activation function used is 'relu'. The input data size is dictated by the size of the spectrogram image used as the feature. The architecture of the synthesized convolutional neural network is shown in Fig. 4.

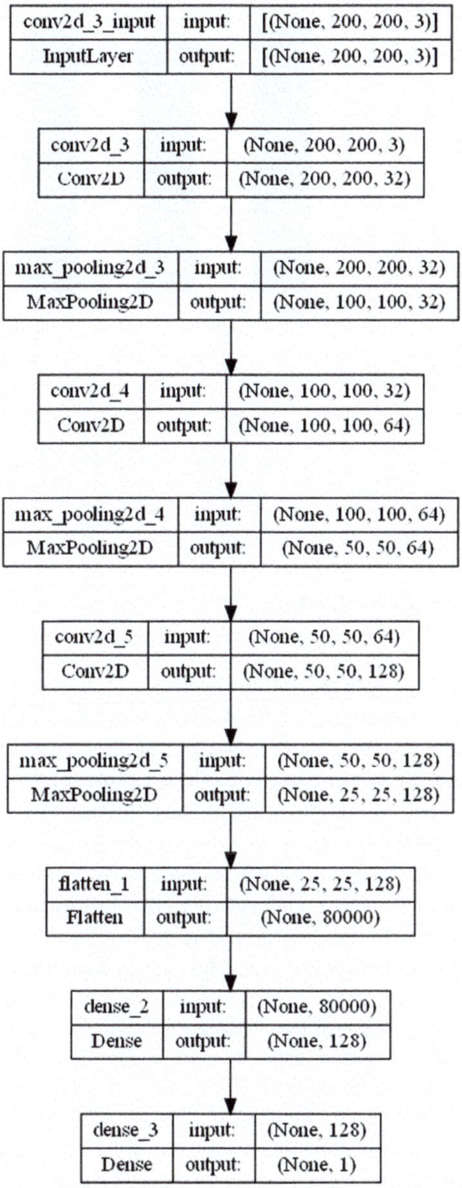

Fig. 4 Architecture of the synthesized CNN

5 Results

The classifiers were compared using the widely used measure of classifier efficiency "F1 score" [4, 5]. The results of the classifiers are presented in Fig. 5. The best result was shown by the LDA classifier (F1 score = 0.91). Similar results were shown by

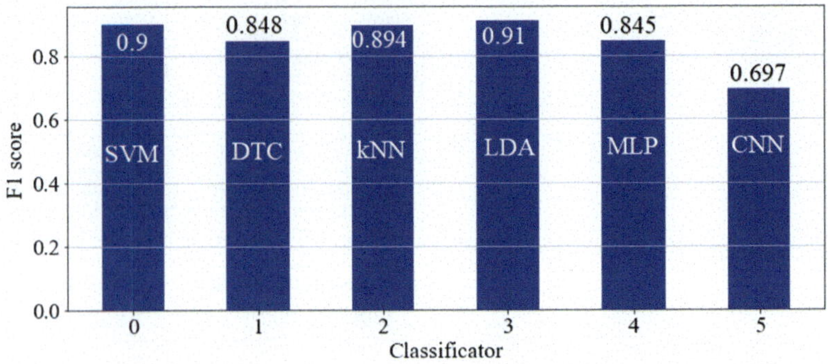

Fig. 5 Classifier results

the SVM (F1 score = 0.9) and kNN (F1 score = 0.894) classifiers. The worst result was shown by the synthesized convolutional neural network (F1 score = 0.697).

6 Conclusion

All classifiers, except the classifier based on a convolutional neural network, showed good results. A comparative analysis of the results obtained allows us to conclude that the binary classification of the movement representation of the left or right hand can be performed with high accuracy, which makes it possible for limited control of robotic devices. Increasing the number of defined classes will further expand this capability.

Although the CNN classifier showed lower F1 score, it achieved it without manual preprocessing in form of CSP filtration. The low result of the convolutional neural network may be due to the low resolution of the spectrograms entering the classifier input. Therefore, this topic requires further research.

Acknowledgements This research was funded by the Ministry of Science and Higher Education of the Russian Federation under the agreement № 075-15-2022-303 to support the development of a World-class research center "Pavlov Center for Integrative Physiology for Medicine, High-tech Healthcare, and Stress Tolerance Technologies"

References

1. Baniqued, P.D.E., Stanyer, E.C., Awais, M., Alazmani, A., Jackson, A.E., Mon-Williams, M.A., Mushtaq, F., Holt, R.J.: Brain–computer interface robotics for hand rehabilitation after stroke: a systematic review. J Neuroengineering Rehabil. **18**(1), 1–25 (2021)

2. Saha, S., Mamun, K.A., Ahmed, K., Mostafa, R., Naik, G.R., Darvishi, S., Khandoker, A.H., Baumert, M.: Progress in brain computer interface: challenges and opportunities. Front. Syst. Neurosci.Neurosci. **15**, 578875 (2021)
3. Värbu, K., Muhammad, N., Muhammad, Y.: Past, present, and future of EEG-based BCI applications. Sensors **22**(9), 3331 (2022)
4. Sinigaglia, C., Rizzolatti, G.: Through the looking glass: self and others. Conscious. Cogn.Cogn. **20**(1), 64–74 (2011)
5. Altuwaijri, G.A., Muhammad, G.: A multibranch of convolutional neural network models for electroencephalogram-based motor imagery classification. Biosensors **12**(1), 22 (2022)
6. Blankertz, B., Dornhege, G., Krauledat, M., Müller, K.R., Curio, G.: The non-invasive Berlin brain–computer interface: fast acquisition of effective performance in untrained subjects. NeuroImage **37**(2), 539–550 (2007)
7. Chen, W., Cai, Y., Li, A., Su, Y., Jiang, K.: EEG feature selection method based on maximum information coefficient and quantum particle swarm. Sci. Rep. **13**(1), 14515 (2023)
8. Ali, M.S., Hassan, A., Rahim, A., Ashraf, M.H., Rahim, A., Saghir, S.: Motor imagery EEG classification using fine-tuned deep convolutional efficient NetB0 model. In: 2023 3rd International Conference on Artificial Intelligence (ICAI), pp. 1–6. IEEE (2023)
9. Saha, P.K., Rahman, M.A., Alam, M.K., Ferdowsi, A., Mollah, M.N.: Common spatial pattern in frequency domain for feature extraction and classification of multichannel EEG signals. SN Comput. Sci. **2**, 1–11 (2021)
10. Ilyas, M.Z., Saad, P., Ahmad, M.I., Ghani, A.R.I.: Classification of EEG signals for brain-computer interface applications: Performance comparison. In: 2016 International Conference on Robotics, Automation and Sciences (ICORAS), pp. 1–4. IEEE (2016)
11. Tang, X., Wang, T., Du, Y., Dai, Y.: Motor imagery EEG recognition with KNN-based smooth auto-encoder. Artif. Intell. Med.. Intell. Med. **101**, 101747 (2019)
12. Antony, M.J., Sankaralingam, B.P., Mahendran, R.K., Gardezi, A.A., Shafig, M., Choi, J.G., Hamam, H.: Classification of EEG using adaptive SVM classifier with CSP and online recursive independent component analysis. Sensors **22**(19), 7596 (2022)
13. Alotaiby, T.N., Alshebeili, S.A., Alotaibi, F.M., Alrshoud, S.R.: Epileptic seizure prediction using CSP and LDA for scalp EEG signals. Comput. Intel. Neurosci. (2017)
14. Gaurav, G., Anand, R.S., Kumar, V.: EEG based cognitive task classification using multifractal detrended fluctuation analysis. Cogn. Neurodyn.. Neurodyn. **15**(6), 999–1013 (2021)
15. Hramov, A.E., Maksimenko, V.A., Pchelintseva, S.V., Runnova, A.E., Grubov, V.V., Musatov, V.Y., Zhuravlev, M.O., Koronovskii, A.A., Pisarchik, A.N.: Classifying the perceptual interpretations of a bistable image using EEG and artificial neural networks. Front. Neurosci.Neurosci. **11**, 305098 (2017)
16. Li, Y.: CNN-based image analysis for EEG signal characterization. Adv. Biomed. Bioinform. Eng. IOS Press, 20–30 (2023)

SPRINGER NATURE

GPSR Compliance

The European Union's (EU) General Product Safety Regulation (GPSR) is a set of rules that requires consumer products to be safe and our obligations to ensure this.

If you have any concerns about our products, you can contact us on ProductSafety@springernature.com

In case Publisher is established outside the EU, the EU authorized representative is:

Springer Nature Customer Service Center GmbH
Europaplatz 3
69115 Heidelberg, Germany

The manufacturer's authorised representative in the EU is Springer Nature Customer Service Centre GmbH, Europaplatz 3, 69115 Heidelberg, Germany. If you have any concerns regarding our products, please contact ProductSafety@springernature.com

Printed and bound by CPI Group (UK) Ltd, Croydon, CR0 4YY
01/10/2026
02225969-0008